LIGHTFOOT GUIL

~

WILD FOODS BY THE WAYSIDE

Heiko Vermeulen

ISBN: 978-2-917183-23-6
Revision 1

Also by LightFoot Guides

Riding the Milky Way 2006
Riding the Roman Way 2007
Reflections - A Pictorial Journey Along the via Francigena 2008
LightFoot Guide to the via Francigena - Canterbury to Besançon 2008/9/10/12
LightFoot Guide to the via Francigena - Besançon to Vercelli 2008/9/10/12
Companion to the via Francigena 2010/2011
LightFoot Guide to the via Domitia 2011
LightFoot Companion to the via Domitia 2011
LightFoot Guide to the Three Saints Way - Winchester to Mont St Michel 2008
LightFoot Guide to the Three Saints Way - Mont St Michel to St Jean d'Angely 2008

Acknowledgements

Thanks go first and foremost to my long suffering wife Susan for her patience over the years as time and again I dived into yet another undergrowth off the beaten track to look for some elusive plant or other and for her willingness to try my various exotic concoctions. Secondly thanks goes to Babette of Pilgrimage Publications for her unwavering support, proof reading, editing not to mention friendship. Of the many people who showed me one plant or another or inspired a recipe I would like to highlight Remigio Pagliari, president of the Gruppo Micologico Naturalistico del Club Alpino Italiano – Sezione Sarzana, for being a fountain of knowledge about the local edible flora. Finally a general thank you to all my blogger friends all over the world who have followed my ramblings, commented on them, sometimes helping identifying plants and for implanting the seed which has become this book. Many of you, I know, prefer to stay anonymous.

Photo Credits go to Andrea Miani (author photo) Nicole Ruskelle (Asparagus and Caper), Tom Connally (Asparagus), Jenn Murphy (Bracken), Teleri Williams (Bugloss, Italian), Sean Rowe (Burdock), Mike Hannon (Clover, Echinacea, Salad Burnet). Thank you for letting me use your photos.

Disclaimer
Care has been taken to ensure all plants in this book have been identified correctly and that all the information is accurate. However I am not a botanist, but a self-taught wild food forager. I cannot take any responsibility for any adverse effects from the use of plants either as food plants or medicinally. Always seek advice from a professional before using a plant medicinally and only consume plants you have positively identified. If you spot any mistakes or omissions within the text, please get in contact and I will make sure it will be corrected in future editions.

Introduction

Wild food foraging has been a bit of a buzz term recently and has enjoyed something of a renaissance. A restaurant in Denmark has won the World's Best Restaurant Award two years running with a menu of local and foraged food. According to some cynics in the press this is all a passing fad, but some of us have been foraging for years and I believe we are simply following an inner urge, something that is in our collective DNA, and the very brief hiatus in human history that we have lost the ability to gather foods in the fields and woods is the passing fad. Only as recently as the generation of my parents, people made it through the war by foraging foods and in countries like Italy the older generation still are seen to dive into grass banks for their daily meals and the younger generation are keen to follow their lead.

The shortages of World War II are still a living memory in many parts of Europe and the wisdom is that people who own a small plot of land and know how to find sustenance from spontaneously appearing plants can withstand anything a war, an economic crisis or an unpredictable government can throw at them. I have been foraging on and off for the last 40 years, initially out of a general interest in nature as a kid and later out of necessity when I was simply too poor to buy food. In these times of economic instability foraging is becoming an ever more important skill to possess.

Nowadays if I look at a meadow I think lunch. If a complete list of all species of plants was to be compiled and you were to split the list into edible and non-edible, I am convinced the edible column would be the longer one. If you then add plants with medicinal properties and other uses to humans to the edible list, what is left over would fit onto the back of a postage stamp. Hence this book makes no claim to be a complete guide, this would make it much too big to be carried around. I have concentrated on plants easily found on land, completely omitting the complex field of fungi, to which many useful guides exist. This book is aimed at the long distance or casual hiker anywhere in Western Europe. There are some guides to edible plants in English concentrating on the flora of the UK or the USA. Other guides for their respective countries are available in German, French or Italian. I have tried to distil this into a guide to the plants of the greater biodiversity comprising the different climatic zones from the temperate North to the Mediterranean climates of the South.

Why Forage

Foraged food is free. We live in times where we worry about our personal financial security often due to circumstances beyond our control. Why not save money by picking food that nature freely gives us in abundance.

Wild food is healthy. Over the years since humanity invented agriculture we have bred cultivated varieties of wild plants to become bigger, more productive, more resistant to pests and diseases, brighter coloured, straighter shaped, occasionally for taste. What has been neglected was actual nutritious content. The wild cousins of our cultivated plants contain far more nutrients than the fruit and vegetables we buy in our supermarkets. Wild plants are also untreated and therefore organic. Think of the premium you pay for organic produce! You can have organic greens for nothing. Of course there is always a degree of pollution that taints even plants gathered in pristine environments.

Wild food is tasty. Anyone who has ever tasted a wild strawberry compared to a cultivated one knows what I am mean. These tiny pea-sized fruits have ten times the flavour intensity of a whole punnet of supermarket strawberries! Yes it is true that some greens are also that much more bitter than their cultivated counterparts, but it is a matter of weaning ourselves off processed, sugar-rich foods, which do no good to our health and re-adjust our taste buds. By no means all wild greens are bitter and if you use some bitter leaves in combination with other less bitter ingredients, they can actually challenge your palate. I was in the wine trade for many years and I know what I am talking about. A fine Bordeaux is only a fine Bordeaux if it contains quite high levels of bitter tannins. These ensure the wine can age gracefully, after which the tannins will be in balance with the other flavour components of the wine.

Foraging connects us back to nature. This guide is aimed at hikers. How many times have you wondered what this plant is called, where that smell comes from, what use this plant might have? Once you find out that so many of the plants you encounter not only are edible, but delicious and healthy, many of them you have known for years but didn't realise you could eat, it becomes an obsession. Keep your eyes open and you will be surprised what nature wants to give you.

Wild food in an emergency. You have taken that small trail over the mountains. The last village you passed had shut down for lunch and your reserves are running low. To top it you got lost and you are running out of daylight. Your only option is to put up the tent and take a fresh look in the morning. With just a few basics and what can be found around you, you can at least cook yourself a comforting meal and feel as smug as Raymond Mears on an SAS survival mission. There are also many plants which can be used as emergency wound dressing and having other medicinal properties.

Rules to Foraging

Respect nature. Don't trample down a load of other vegetation to get to your favourite morsel. Never graze an area bare. Leave some for other wildlife and give the plants a chance to regenerate themselves, so next year you or someone like you will again be able to gather some. Don't gather protected species.

Respect private property. Many plants can be found on meadows or amongst olive groves, vineyards or fruit orchards which are strictly speaking private property. Most owners will not mind you picking some of their weeds. However do not pick crops which have evidently been planted deliberately and if a habitation is nearby it may be prudent to ask for permission before trespassing on their land.

Respect your own health. Every year some 30 people in Italy die fungi hunting. Only about two of them die from actually ingesting poisonous mushrooms, the rest just wanted to get to that elusive one right on the edge

of that cliff. They slip and injure themselves and it may be days before they are found. Of course ensure you know what you are eating, maybe just testing a small quantity if not 100% sure. Few plants in temperate Europe are deadly, most unsuitable plants will just make you sick. Dangerous animals are also far and few in between, the only one really is the venomous Common European Adder or Viper. Most snakes you encounter are harmless and will scuttle out of your way as you approach. The viper is a much slower snake and will simply stay put and if you do get too close may strike. It can be recognised by its bronze / brown colour and a darker pattern along its back and a relatively stout body. If walking through undergrowth it may therefore be best to disturb the ground in front of you with a long stick and wear boots covering your ankles.

Avoid polluted areas. This really is part of the former point too. It may be self-evident, but avoid picking plants destined for your plate which grow next to busy roads, near industrial installations, along polluted rivers or on or near cultivated ground which may have had chemical herbicides, pesticides or fertilisers applied to them. Unfortunately we are no longer in a position to find completely chemical free food as every new born child even in the remotest arctic or desert regions of this world is born with man-made chemicals in their systems, but it is prudent to avoid the higher concentrations found near the above mentioned sites.

How to identify plants

Sight: Take a careful look at the foliage, stems, flowers and possibly roots of a plant. Also take into consideration its location, whether it grows in the shade or on a sunny location, in association with other plants, on a dry or wet spot, the time of year, whether in a woodland, by a river or along the sea or on an open field and possibly the soil type.

Smell: Many plants exude a distinct aroma when crushed or directly off the flowers. If the aroma is unpleasant the likelihood is that the plant is not edible.

Taste: If you feel you may have a positive identification, nibble just a small amount and spit out again. Many species, especially those of the asteracea family, the dandelion and its many look-alikes, have no poisonous relations, just some taste better than others and taste can safely be used as an identifier.

If still in doubt: Take a sample of the plant and take a note where you found it. I have provided links with more detailed descriptions of plants in each section. Do some more research next time you can access the internet.

In some countries, such as Italy and Germany it is possible to take fungi and wild plants to a pharmacy, where there are experts at hand who can help identify them.

All you need now is a penknife, a sturdy trowel and a collection basket. Happy Foraging! If your path takes you down the Via Francigena, just as you glimpse the Mediterranean for the first time entering Liguria, you are near me. I offer guided wild food forages in the greater La Spezia region given a bit of notice.

Contact me through my blog on: http://pathtoselfsufficiency.blogspot.it/.

Useful web-links

Path to Self Sufficiency:
http://pathtoselfsufficiency.blogspot.it/ My own blog with many references to wild foods as well as gardening and life in Italy.

Plants for a Future data base:
http://www.pfaf.org/user/default.aspx. A detailed account of over 7000 edible and medicinal plants. My first port of call for any queries.

Botanical dot com
A modern Herbal:
http://botanical.com/. A beautifully illustrated guide to medicinal and edible herbs. An electronic version of a 1931 original.

Flora Italiana:
http://luirig.altervista.org/. In Italian but with many useful photos and translations into other languages.

British Wild Flowers:
http://www.british-wild-flowers.co.uk/index%20flowers.htm.

Eat the Weeds:
http://www.eattheweeds.com/. A blog by Green Deane, a forager in the USA

Fergus the Forager:
http://wildmanwildfood.blogspot.it/. Another great blog by experienced forager Fergus in Kent.

Wild Food and Recipes:
http://www.wildfoodandrecipes.co.uk/. Based in the UK

Subsistence Pattern:
http://subsistencepatternfoodgarden.blogspot.it/. An informative blog by my good blogging pal Mr.H in Idaho. He writes about subsistence farming and foraging.

Valle Nuova:
An Italian B&B keeper in Le Marche, blogging in English and often mentioning foraging.

Sacred Earth:
http://www.sacredearth.com/.
Many useful articles on wild foods.

Foodolutions:
http://foodolution.com/blog/. Some
nice ideas on what to do with foraged foods.

Cuisine Campagne:
http://www.cuisine-campagne.com/index.php?. Cookery blog in French by author Linda Louis, who incorporates many wild ingredients in her cooking.

Acta Plantarium:
http://www.actaplantarum.org/index.php. In Italian, but with many useful pictures.

Piante Spontanee in Cucina:
http://www.piantespontaneeincucina.info/struttura/html/piante_spontanee_i n_cucina/piante_spontanee_in_cucina.html. If you speak Italian this is one of the most useful sites describing many different edible species and giving recipes for most in pdf format.

Plants of Spain:
http://www.topwalks.net/plants/index.htm. A multilingual site with many pictures.

The Wild Food Resource:
http://www.topwalks.net/plants/index.htm. From Nature's Secret Larder, this is a comprehensive guide to wild foods in Britain.

Micological Group of Livorno:
http://www.gruppomicologicolivornese.it/Erbe_spontanee_commestibili.ht ml. Pictures of wild herbs as well as fungi in Tuscany

Angelica (*archangelica sylvestris*)

Description: Biennial plant with serrated bright green leaves, hollow stems with celery like texture. Flowers in summer white umbrella-like. Grows up to 2m tall and has a distinct delicately sweet aroma.

Where: Along pathways in shady or semi-shady locations throughout Europe.

When: Spring to summer for leaves and stems, autumn for roots and seeds.

Culinary uses: All parts of the plant are edible. The hollow stems can be crystallised and used as cake decorations. Fresh stems and leaves can be used with stewed fruit such as plums or rhubarb. Seeds are one of the main ingredients to herbal liqueurs such as Chartreuse and also in Vermouths. The roots of young plants can be boiled and eaten as a vegetable similar to parsnip. Leaves can be added sparingly to a salad.

Medicinal uses: A decoction of the roots and seeds is used in the treatment of bronchial catarrh, coughs and dyspepsia.

Caution: May increase skin's sensitivity to sunlight and lead to dermatitis. Should also be avoided by people with tendency towards diabetes.

Recipe: Crystallised Angelica Stems: Take about finger-thick pieces of stem and cut into 5 cm lengths. Boil until tender then peel thick outer skin off. Weigh and add the same weight in sugar. Leave covered in a bowl for a few days. Boil again for a few minutes, drain, weigh and add same amount of sugar again. Leave covered for another 2 days, then dry in a food dehydrator until crisp. Use as cake decoration or as sweets for the kids.

Links: http://www.botanical.com/botanical/mgmh/a/anegl037.html
www.pfaf.org/user/Plant.aspx?LatinName=Angelica+sylvestris

Apple / Crab Apple (*malus sylvestris*)

Description: The apple and it's wild cousin the crab apple don't need much of a description.

Where: I do not suggest you stealing from orchards, but often evidently abandoned orchards can be found. If the ground is overgrown and fruit lies rotting on the ground, it is safe to assume that no one will mind you picking the fruit. If a habitation is nearby it may be best to ask. Crab apples are found in some parklands or mixed deciduous woods.

When: Autumn

Culinary uses: No need to describe the cultivated apple. Crab apples are

high in pectin and therefore make great additions to other fruits, notably blackberries, in the making of jams and jellies, instead of adding commercial pectin.

Medicinal uses:
Crushed pulp of crab apples can be used as a poultice for skin inflammations and small flesh wounds. The fruit is also laxative.

Caution: The seeds of apples contain the toxin hydrogen cyanide and should not be consumed in large quantities.

Archangel, Yellow (*lamium galeobdolon*)

Description: A perennial member of the dead nettle family this has small nettle-like leaves similar to those of red dead nettle, but tinged yellow towards the top. The flowers are bright yellow and tubular in shape. It grows up to 40cm in height and often makes a ground cover. The stems are distinctly angular.

Where: On fields and shady moist places throughout Europe.

When: Spring to summer for leaves and flowering shoots.

Culinary uses: The young leaves, shoots and flowering tips can be cooked as a pot-herb.

Medicinal uses: The plant makes a good wound herb, applied to small cuts.

Links: http://www.pfaf.org/user/Plant.aspx

Asparagus, Wild (*asparagus officinalis*)

Description: Woody ferny weed growing up to 1.80m, sprouting thin green shoots in spring which eventually bear small red berries. The shoots

themselves can grow up to 1.5m. Wild asparagus is much thinner than the cultivated variety. It takes some practice to spot the shoots when they arrive. Look out for the ferny weeds first and then carefully scan the general area around it. Shoots can sometimes appear quite a distance from the actual plant. It is easily confused with hop shoots or the shoots of Black Bryony, both of which grow in similar locations and are just as edible and tasty. The head of the hop shoot is a little more open compared to that of asparagus and the stems are slightly square and rough to touch. Those of Black Bryony grow longer and wind themselves around surrounding plants and they form little heart-shaped leaves.

Where: Mixed woodlands

When: Early spring

Culinary uses: Used like the cultivated variety, lightly steamed with some butter, lemon juice and salt, as an ingredient for a frittata or a soup. They are a good source of protein and dietary fibre as well as phosphorus, zinc, vitamins A and C.

Medicinal uses: The shoots have a cleansing effect on the bowels, kidneys and liver. An infusion made from the root (harvested in late spring after the shoots have been picked) is useful in the treatment of many urinary problems including cystitis and is said to be useful in the treatment of cancer.

Caution: Consumption of asparagus gives the urine a strongly unpleasant smell. Large quantities may irritate the kidneys.

Recipe: Asparagus Risotto: Briefly blanch asparagus in boiling water, drain and roughly chop. Fry an onion in some butter until soft, add the asparagus then some risotto rice. After a minute add a glass of white wine and cook until the wine is evaporated. Bit by bit add chicken stock until all liquid is absorbed and the rice is al dente. Season to taste and serve with some Parmesan cheese.

Links:
http://www.pfaf.org/user/Plant.aspx?LatinName=Asparagus+officinalis

Autumn Olive Tree (*elaeagnus umbellate*)

Description: This deciduous shrub or low tree originated from Eastern Asia and was introduced to Europe and North America to plant alongside river banks to prevent erosion. It has proved so successful that it is considered an invasive species in some places. It has nothing to do with the real olive tree. The name relates to the underside of the foliage which is coloured silvery like the leaves of an olive tree. In late autumn it carries oodles of bright red fruit with some silvery mottles growing in little clusters not unlike red currants. Alternative names are Autumn Berry, Oleaster, Silver Berry and Aki-Gumi.

Where: Alongside river and canal banks.

When: Fruit ripens late autumn

Culinary uses: The fruit eaten raw is in taste somewhere between redcurrants and cranberries. It can also be dried to add to cereals or made into jams.

Medicinal uses: The fruit is one of the best known sources of Lycopene; about 16 times that of tomatoes. Lycopene is a powerful anti-oxidant and therefore a great preventative for heart disease and cancer. It also contains high levels of vitamins A, C and E as well as flavonoids and essential fatty acids.

Recipe: Christmas Jam: 500g of autumn olive berries, 500g of strawberry tree fruit 150g of myrtle berries, 1 chopped apple, zest & juice of 1 orange, cinnamon, vanilla, ground nutmeg, ginger & cloves. Cook until soft, add 500g sugar and boil fast until setting point. All these fruits are ripe around the same time; just before Christmas.

Link: http://davesgarden.com/guides/articles/view/1758/

Bay Tree (*laurus nobilis*)

Description: Large evergreen tree with hard, glossy leaves exuding a pleasant aroma when rubbed.

Where: Common in mixed woodlands throughout the Mediterranean

When: All year

Culinary uses: One or two leaves used in stews and meat dishes and removed before serving impart a pungently aromatic flavour. They are also part of the 'Bouquet Garni' herb mixture alongside parsley, marjoram, thyme and some peppercorns. The dried leaves can be used to make a tea.

Medicinal uses: It's known as an aid to digestion, an appetite stimulant and is used in the treatment of bronchitis and influenza. It is also said to be effective against various types of cancer. Dried leaves are also effective as insect repellents.

Recipe: Scamorza on Bay Leaves: Place a piece of smoked Scamorza cheese or other smoked cheese on a bay leaf. Make as many as required, place on a baking tray and heat in the oven until cheese is melted. Serve as an appetiser, just sucking the aromatised melted cheese off the leaves.

Link:
http://www.pfaf.org/user/Plant.aspx?LatinName=Laurus+nobilis

Beech Tree (*fagus sylvatica*)

Description: Large deciduous tree with a smooth bark, common throughout temperate regions of Europe. The leaves are a simple oval shape and the fruit develops in soft spiky hulls which eventually spill their little triangular nuts.

Where: Mixed woodlands often with oak or pine trees. More common north of the Alps

When: Young leaves in spring, nuts in autumn

Culinary uses: The very young leaves make a pleasant addition to the salad bowl. The nuts are very fiddly to extract, but taste delicious, eaten raw or roasted sprinkled over salads or nibbled as a snack. A good quality salad oil can be extracted from the seeds. You have to be quick though before the squirrels get to them.

Medicinal uses: Not really used for self-medication. The bark and a tar made by distilling branches are used for certain conditions.

Caution: Large quantities of beechnuts may be poisonous.

Link: http://www.pfaf.org/user/Plant.aspx?LatinName=Fagus+sylvatica

Beet, Wild (*beta vulgaris*)

Description: A biennial plant common in Southern Europe, this is a relation to Swiss chard and beetroot. The leaves and stalks are similar to Swiss chard, but generally smaller and more delicate. Tiny flowers are produced on dense spikes. A variant, Sea Beet, is found near the coast. In fact most wild beet is probably a cross between Sea Beet and cultivated Chard.

Where: On open ground near paths and on cultivated land.

When: Summer

Culinary uses: Stalks and leaves can be cooked like spinach. They are a popular original ingredient in many versions of vegetable tart all over Italy.

Medicinal uses: A decoction made from the seeds is used in the treatment of intestinal and genital tumours. The juice of other parts of the plant is used in treating many types of cancer and Leukaemia. Beet juice in vinegar gets rid of dandruff on the scalp. The juice can also be applied to ulcers. Juice applied inside the nostrils is said to alleviate ringing in the ears and toothache.

Recipe: Scherpada (vegetable tart of Ponzano Superiore) My favourite recipe of la Cucina Povera, the poor cuisine of Italy. Every last weekend in August the village of Ponzano Superiore on the Via Francigena holds a 4 day festival dedicated to this speciality:

Ingredients for the filling: olive oil, wild beet, wild leek, squash, breadcrumbs, grated pecorino cheese, salt & pepper.

Ingredients for the pastry: flour, salt & water
Method: Sauté the vegetables together until soft. Season well and add cheese and breadcrumbs to give a reasonably solid mass. Combine flour, salt and water and knead until a smooth but not sticky consistency is reached. Cut into 2 halves and roll each out as thinly as possible. Line a pie tin with one half, fill with the filling and cover with the other half of the pastry. Brush a little oil on top and bake for about 45 minutes until nicely brown. Sprinkle a little more oil and cheese on top and serve.

Link:
http://www.pfaf.org/user/Plant.aspx?LatinName=Beta+vulgaris+maritima
http://en.wikipedia.org/wiki/Beets

Bellflower (*campanula*)

Description: A large family of flowering plants characterised by their bell-shaped flowers. There are some 500 sub-species, so needless to say it would be beyond this guide to describe them in detail. Suffice to say that all of them to a lesser or greater extend have edible leaves, all have good tasting flowers and for some the roots are also edible. Most common are the Large Bellflower, c. latifolia, pictured above left and Rampion Bellflower, c. rapunculus, pictured centre and right. The former grows up to 1.5 metres tall and has up to 10cm long, bright blue flowers, the latter grows up to 1m tall with more delicate pinkish to light blue flowers and edible roots.

Where: Large campanula common on woodland edges and along hedgerows. Smaller species are often found in large clusters in woodland clearings or rock crevices.

When: Summer

Culinary uses: The flowers make a sweet tasting and decorative addition to a salad. Leaves of some species are also tasty. None are poisonous, so simply taste and see if you like them. Young shoots in spring can be eaten raw or cooked and contain high amounts of vitamin C. The roots of some species (campanula rapunculus) are edible, and can be added raw to salads to add some crunchiness.

Medicinal uses: None known

Link: http://en.wikipedia.org/wiki/Campanula

Birch, Silver (*betula pendula*)

Description: Small to medium deciduous tree with distinct white bark and fine, heart shape foliage

Where: Common in Northern Europe, in Southern Europe restricted to higher altitudes.

When: Spring

Culinary uses: A country wine can be made from the young leaves or the sap. I haven't tried this, but the inner bark can be ground into a meal as thickener for soups or mixed with flour to make bread. The sap also makes a refreshing drink. Young leaves can be eaten cooked or raw.

Medicinal uses: Helps alleviate bladder and kidney problems. A tea made from 2 tsp of dried young leaves per cup can be used as a spring cleanser. It is also used in the treatment of gout, dropsy and rheumatism, and is recommended as a reliable solvent of kidney stones.

Link:
http://www.pfaf.org/user/Plant.aspx?LatinName=Betula+pendula

Bistort (*polygonum bistorta*)

Description: This perennial herbaceous flowering plant is not one of my favourites, but gets included for its traditional role in Northern England to make Easter-Ledge Puddings. It gets it's name from the twisted twin root from which large, long simple dock-like leaves emerge and in the summer pretty candle-like pink flowers.

Where: Damp meadows and along waterways all over Northern and Central Europe. In the South it is restricted to mountainous areas.

When: Late winter to early summer for the leaves.

Culinary uses: The leaves are used in a traditional Lenten pudding in many parts of Northern England (see recipe). They can be used as a potherb instead of spinach, although contrary to other reports, I find them bitter. The roots can be eaten but need to be boiled and roasted to make them palatable.

Medicinal uses: Infuse 2 tsp of chopped root a few hours in lukewarm water and take against diarrhoea. This can also be gargled to reduce mouth ulcers and infections of the throat. It is also effective against internal and external bleeding and reduces excessive menstrual bleeding.

Recipe: Easter Ledges Pudding: Boil a good handful of washed and chopped bistort leaves plus a few nettle tops for 10 minutes, then strain. Add 1 beaten raw egg and one finely chopped hard-boiled egg. Season with salt & pepper and mix in a little soft butter. Boil in a pudding basin covered with cloth for 40 minutes. Serve with veal around Easter time.

Caution: Consume in moderation as it contains oxalic acid which can inhibit the bodies ability to absorb other vital vitamins. It may also cause photosensitivity in some people.

Links:
http://www.pfaf.org/user/Plant.aspx?LatinName=Polygonum+bistorta

Bittercress, Hairy (*cardamine hirsuta*)

Description: Also known as Lamb's Cress, this is a low growing annual plant with a rosette of deeply lobed shiny leaves, from which a wiry stem emerges with small white flowers.

Where: An invasive weed on cultivated ground, on damp, disturbed ground and in wall crevices.

When: Late autumn through to early spring.

Culinary uses: The leaves have a hot cress-like flavour and can be added to salads or cooked as a potherb.

Medicinal uses: Rich in Vitamin C and stimulates the appetite.

Recipe: Bittercress soup: Gently fry 1 sliced leek and a few cubed potatoes in some oil. Add 1 litre of water and simmer until potatoes are cooked. Season to taste and stir in 7 chopped rosettes of bittercress and some crème fraîche. Serve with a crust of bread.

Links:
http://www.pfaf.org/user/Plant.aspx?LatinName=Cardamine+hirsuta

Blackberry (*rubus fruticosus*)

Description: A vigorous perennial, deciduous shrub with thorny stalks and leaves producing an abundance of black clustered berries.

Where: Woodland edges, hedgerows, scrubland.

When: From late Summer into Autumn, depending on location and variety.

Culinary uses: The fruit is delicious raw or cooked and can be used in all manner of ways. It has high nutritional contents of dietary fibre, vitamin C, vitamin K, folic acid and the essential mineral manganese. It is also antioxidant. The young leaves can be dried and made into tea. Young shoots coming out of the ground can be peeled and added raw to salads, although in my experience they are somewhat stringy and peeling them is fiddly.

Medicinal uses: The root bark and leaves make an excellent remedy for dysentery, diarrhoea, haemorrhoids and cystitis. Externally, they are used as a gargle to treat sore throats, mouth ulcers and gum inflammations. A decoction of the leaves is useful as a gargle in treating thrush and also makes a good general mouthwash.

Link:
http://en.wikipedia.org/wiki/Blackberry

Blueberry (*vaccinum myrtillus*)

Description: A variety of shrubs, also known as Bilberry, ranging from low growing species of 30 cm to tall 4 m shrubs. The berries are purplish blue with a flared crown at the end.

Where: On acid soils in temperate pine woodlands. In the south restricted to mountainous regions.

When: Summer

Culinary uses: Fruit can be eaten raw or cooked. Leaves are used as a tea.

Medicinal uses: Leaves and stems are used to treat various complaints of the female reproductive system. It is said to be contraceptive and to bring on a delayed period as well as slow excessive period bleeding.

Link:
http://www.pfaf.org/user/Plant.aspx?LatinName=Vaccinium+myrtilloides

Borage (*borago officinalis*)

Description: Annual herb, growing up to 40cm tall with large, prickly, fleshy leaves and beautiful bright blue, drooping flowers. All parts of the plant have a distinct cucumber taste to them

Where: Alongside paths and on open cultivated fields and vineyards.

When: Spring to Autumn

Culinary uses: Leaves finely chopped as salad ingredient. The flowers make a colourful addition to a salad or are added to a refreshing drink. Flowers and leaves make an excellent stuffing for home made ravioli (see recipe). The leaves dipped in batter and fried make a tasty snack. A pleasant tea can be made from the leaves and flowers.

Medicinal uses: The plant is rich in potassium, calcium, mineral acids and a very beneficial saline mucilage. It is beneficial to the circulation of the blood. Compresses made from borage help relieve congested veins in the legs. The tea also helps against rheumatism, cleanses the blood and relieves sore throats. It is believed to lift the spirits too.

Caution: People with liver problems should refrain from eating large quantities of Borage.

Recipe: Borage Ravioli: For the stuffing combine 500ml of borage leaves, 250g ricotta cheese, juice of ½ lemon, 100g grated parmesan, 2 tbsp breadcrumbs, pepper and 2 tbsp olive oil in a food processor. For the dough combine 100g white flour per person, 1 egg, a pinch of salt and 1 tbsp of oil. Roll out thinly and cut into rectangles. Stuff and seal into ravioli shapes. Cook a few minutes in boiling water until they rise to the surface. Serve with some sage fried in butter and a sprinkling of parmesan.

Link:
http://www.pfaf.org/user/Plant.aspx?LatinName=Borago+officinalis

Bracken (*pteridium aquilinum*)

Description: This common coarse species of fern apparently gained its Latin name aquilinum (eagle-like) due to the fact that with a lot of imagination you can see a German Imperial Eagle in the surface of a cut stem. This fact has been bestowed on me by famous botanist David Bellamy.

Where: In woodlands all almost all over the world

When: Spring

Culinary uses: All northern hemisphere ferns are edible in the 'fiddlehead' stage. They are best cooked. I fry them in butter with a little lemon juice and served as an appetizer. According to some sources the dried and powdered root can be used to make dumplings. It is high in starch content but may cause constipation.

Medicinal uses: The young shoots are diuretic, refrigerant and vermifuge. Leaves are used in a steam bath to treat arthritis. A tincture of the root in wine is used in the treatment of rheumatism. A tea made from the roots is useful to combat many ailments including stomach cramps, diarrhoea, to expel worms, internal bleeding, colds and chest pains.

Caution: Bracken is implicated in possibly causing stomach cancer. Therefore only small quantities should be taken and not on a too regular basis.

Link:
http://www.pfaf.org/user/Plant.aspx?LatinName=Pteridium+aquilinum

Broom, Spanish (*spartium junceum*)

Description: This up to 3.5m tall perennial shrub is mentioned chiefly to distinguish it from gorse and for its other use for foragers, namely to make your own collecting basket on the go from the rush-like twigs. The plant looks very similar to gorse with its bright yellow flowers, but in contrast to gorse it has no spikes and the flowers are not edible.

Where: Sunny locations on lime soil, often amongst rocks, mostly in southern Europe, but also naturalised further north

When: Flowering shoots are harvested in spring.

Culinary uses: None

Medicinal uses: The flowering shoots used fresh or dried are strongly diuretic and purgative as well as acting as a heart tonic.

Caution: Its medicinal properties are very powerful and large quantities can cause stomach upsets, diarrhoea and vomiting.

Link:
http://www.pfaf.org/user/Plant.aspx?LatinName=Spartium+junceum

Bryony, Black (*tamus communis*)

Description: A perennial climbing plant growing up to 3.5 metres long, winding itself around any other nearby plant. Initially in spring asparagus like shoots appear, often in the same habitats as asparagus itself, adding to the confusion. Soon shiny heart-shaped leaves unfold. The flowers are small, greenish and inconspicuous which later give way to bright red berries of around 1cm diameter.

Where: Mixed woodlands and hedge banks in semi-shaded positions all over Europe, but most common from Italy eastwards.

When: Shoots in early spring

Culinary uses: Only the young shoots of this plant are edible, eaten just like asparagus. It is best to change the water once during cooking to remove some of the saponins present (see caution). Try them in any way you would use asparagus or hop shoots. I personally prefer them to true asparagus.

Medicinal uses: Used externally, the macerated root can be applied as a poultice to bruises or rheumatic joints. It may cause irritations though and should only be administered by a qualified herbalist

Caution: The whole plant is poisonous. It contains saponins, which however are common in other foods such as many kinds of beans and are badly absorbed by the body and harmless in reasonable quantities. The plant also contains calcium oxalate crystals, which are an irritant poison and can cause vomiting. These are most concentrated in the berries, which should be avoided at all cost. Only the young shoots make good eating.

Link:
http://www.botanical.com/botanical/mgmh/b/brybla75.html

Bugle, Blue (*ajuga reptans*)

Description: A small perennial flowering plant. The bright blue, upright flowers, growing up to 25cm tall, look a little like the flowers of wild sage, however, the small, oval, dark-green leaves are much smaller than those of wild sage.

Where: Meadows and fields

When: Spring

Culinary uses: The young shoots in spring can be eaten raw, but are somewhat bitter. According to one source the flowers are also edible but according to another they are hallucinogenic and toxic.

Medicinal uses: Bugle has a history as a wound herb applied externally. It also contains the same substance as foxglove making it potentially a remedy for heart disease, however this should only be administered by a qualified herbalist.

Caution: The plant is said to contain a narcotic hallucinogenic. Whilst some people might find that interesting, fatalities have been reported. Only consume young shoots and in moderation.

Links:
http://www.pfaf.org/user/Plant.aspx?LatinName=Ajuga+reptans

Bugloss, Italian (*anchusa azurea*)

Description: A perennial member of the borage family. It looks very similar to a small borage, or a little like forget-me-not, growing up to 0.5m in height with bristly stems and leaves and bright blue flowers. In contrast to borage, the leaves are more narrow and spear-shaped and the flowers have 5 distinct rounded petals.

Where: Arable land, waste grounds and along pathways

When: Spring

Culinary uses: The flowers can be used in the same way as borage flowers as a decorative and tasty addition to a salad. It also compliments fruit salads well. The young leaves and stems can be boiled, steamed or fried as a vegetable.

Medicinal uses: Like borage, bugloss is said to drive away melancholy and depression. Steep a handful of flowers in a litre of red wine for a week, then strain and add 200g sugar for a cheering drink.

Links:
http://www.pfaf.org/user/Plant.aspx?LatinName=Anchusa+azurea

Burdock (*arctium lappa*)

Description: A biennial plant in the thistle family, although not a true thistle as the leaves and stems are not spiny. It is a big conspicuous plant growing up to 1.5 metres with thick leaves reminiscent of rhubarb leaves, but with a downy underside. The flowers look like little purple shaving brushes and once mature the fruit will attach itself to clothing and animal fur.

Where: Waste grounds and along pathways

When: Spring to summer of the second year.

Culinary uses: The most used part is the root, which can be up to 1.2m long. However they are best harvested when they are no longer than about 0.5m as they tend to get woody when older. Unless very young the root will need to be cooked in water for some 20 minutes until tender. It has a mild flavour, slightly sweet and bitter, but it absorbs other flavours as it is cooked. It is rich in proteins and carbohydrates as well as Vitamin C. The flowering stalks that appear in the second year can be peeled, blanched, then cooked until tender and eaten in the same way as cardoons or celery stalks. The immature flower heads can be prepared and eaten in the same way as artichokes.

Medicinal uses: This plant has too many uses to list here. Mostly the dried roots, but also the leaves and fruit are used in both western and Chinese herbalism as a blood purifier. It is used in the treatment of tonsillitis and other infections, against skin rashes and boils and rheumatism. A decoction is made from 30g of root and fruit to ¼ litre of  water and 1 wineglass is taken 3 times a day. The leaves taken as an infusion are helpful for stomach disorders. Externally the leaves can be applied as a poultice to tumours and gouty swellings. The whole plant is antibacterial, antifungal and carminative.

Recipe: Parsleyed Burdock Roots: Peel & slice burdock roots and boil until tender. Drain, return to pan with some butter, salt & pepper, chopped parsley and a splash of lemon juice. Toss and serve hot as a side dish.

Links:
http://outdooredibles.com/burdock/
http://www.botanical.com/botanical/mgmh/b/burdoc87.html

Butcher's Broom (*rucus aculeatus*)

Description: Low evergreen shrub up to 1 m high which looks a little like holly, although the leaves are smaller and less prickly, but the red berries in the autumn and winter look similar and they are equally taken into houses for Christmas decorations.

Where: In woods in shady or semi-shaded positions around the Mediterranean and Western Europe.

When: Young shoots are edible in spring, Roots for medicinal purposes can be dug up any time.

Culinary uses: Young shoots as they break through the soil can be cooked and eaten like asparagus, but is not particularly recommended due to it's rather bitter taste. The seeds can be roasted and ground as a coffee substitute.

Medicinal uses: The dried roots can be made into a tea for the treatment of varicose veins and haemorrhoids.

Caution: The berries are purgative.

Link:
http://www.pfaf.org/user/Plant.aspx?LatinName=Ruscus+aculeatus

Campion, Bladder / Campion, White *(silene vulgaris / s. latifolia)*

Description: An up to 60cm tall perennial flowering plant, with thin branching stalks on which alternate pairs of waxy, greyish-green pointy leaves grow. The flowers consist of 5 divided white petals and the calyx is bloated giving it its common name. White campion is a little less obviously bloated and has smaller leaves, but can be used in the same way as bladder campion. It is particularly popular as a food plant in Spain and Cyprus.

Where: Grassy banks and meadows. Common all over Europe

When: Spring to summer

Culinary uses: The young shoots and leaves can be added raw to salads, adding a mild pea flavour. Older leaves are best cooked and added to soups or vegetable tarts. Pureed they make a great spinach substitute.

They are also good fried and/or added to omelettes. In Spain it is added to rice dishes or soups.

Medicinal uses: Used little in herbalism

Recipe: Arroz con Collejas: Pick a good couple of handfuls of campion leaves, wash well and chop. Fry some cod fillet in olive oil until done. Remove cod from pan and fry some garlic, dried pimento and some chopped tomatoes in the same oil. Once cooked add the campion leaves and sauté for a few more minutes. Then add some rice and stir into the cooking juices. Finally add double the quantity of water to rice and bring to the boil. Once boiling add the cooked cod, some saffron and season to taste. Cook until rice is tender.

Link: http://www.pfaf.org/user/Plant.aspx?LatinName=Silene+vulgaris

Caper (*capperis spinosa*)

Description: A low deciduous perennial shrub, growing up to 1m tall with fleshy round leaves and showy whitish-pink flowers which exude an aromatic perfume.

Where: Hot locations along the Mediterranean, on rocky slopes and on south and west facing walls.

When: The immature flower buds in early summer and the fruit in late summer. Autumn for root bark for medicinal purposes.

Culinary uses: The immature flower buds and fruit is generally pickled and later used as a condiment. Young shoots can be eaten like asparagus.

Medicinal uses: The root bark is used in the treatment of gastrointestinal infections, diarrhoea, gout and rheumatism. It may also have cancer-preventative properties. The flower buds taken internally relieve coughs and externally are applied to eye infections.

Link:
http://www.pfaf.org/user/Plant.aspx?LatinName=Capparis+spinosa

Carrot, Wild (*daucus carota*)

Description: A biennial or perennial umbelliferous plant, which is related to the cultivated carrot. The umbrella of flowers usually starts pink and later turns white. The leaves are fine and resemble those of wild parsley. The long thin root is usually white and smells distinctly of carrot

Where: Waste grounds, on cultivated land and on meadows all over Europe

When: Spring for foliage and flowers, summer for roots and seeds.

Culinary uses: The root, whilst carrot flavoured, is thin and stringy. It's best finely chopped and added as flavouring to a soup. The flowers can be covered in batter and deep fried for a carrot flavoured snack. The foliage can be added to salads or soups and seeds can also be used as flavouring.

Medicinal uses: Wild carrots are rich in vitamins. They act as a mild diuretic and an infusion of all parts of the plant help in the treatment and prevention of cystitis and kidney stones. Wild carrot is also reputed to help with a variety of menstrual problems.

Links:
http://www.pfaf.org/user/Plant.aspx?LatinName=Daucus+carota

Cat's Ear (*hypochoeris radicata*)

Description: Also known as False Dandelion due to it's similarity to this plant. However in contrast to dandelion the leaves are a little fleshier and have a soft furry feel to them. Also the leaves are rather less bitter than dandelion.

Where: Meadows and lawns, waste grounds and open woodlands.

When: Winter to spring

Culinary uses: The young leaves make an acceptable salad ingredient. They can also be boiled or steamed as a potherb. The roots can be dried and roasted to make a coffee substitute.

Medicinal uses: None known

Links:
http://www.pfaf.org/user/Plant.aspx?LatinName=Hypochoeris+radicata

Chamomile, German (*matricaria recutita*)
Chamomile, English or Roman (*anthemis nobilis*)

Description: Amongst the many types of chamomile these two species are the most effective. It is low growing with fine feathery foliage and small daisy-like flowers. When rubbed they will exude the characteristic smell likened to fresh green apples. English Chamomile has a low matting habit, whilst German chamomile has a more upright bushy growth up to 50 cm in height.

Where: Open waste grounds and around cultivated ground in sunny or semi-shaded positions.

When: Summer

Culinary uses: As a tea. Also try a refreshing summer drink half cold chamomile tea sweetened with a little honey, half sparkling water over ice with a slice of lemon.

Medicinal uses: Chamomile tea helps calm frayed nerves, has a soothing effect on the digestion, eases menstrual pain, relieves fevers and colds and is generally relaxing. Externally, added to a hot bath, it will reduce muscular weariness and fatigue.

Link:
http://www.altnature.com/gallery/chamomile.htm

Cherry (*prunus avium & prunus cerasus*)

Description: A medium sized deciduous tree with distinct red berries containing one large stone in the middle. There are two distinct varieties of wild cherry, prunus avium, the sweet cherry and prunus cerasus, the sour cherry.

Where: Mixed woods and abandoned orchards.

When: Early Summer

Culinary uses: Sweet cherries are better for raw consumption. Both can be stewed or turned to jam. Because of it's low pectin content pectin needs to be added for jam making for it to set. I tend to add lemons rather than commercial pectin. Liqueurs are made from the leaves and stones. The latter has a distinct marzipan flavour.

Medicinal uses: The fruit is a rich source of anti-oxidants and vitamin C.

Caution: The leaves and stones do contain some hydrogen cyanide which is toxic. However the concentrations are so low as to be harmless.

Link:
http://www.pfaf.org/user/Plant.aspx?LatinName=Prunus+avium

Chestnut, Sweet (*castanea sativa*)

Description: A large deciduous tree with oblong-lanceolate, boldly toothed large leaves. In autumn it bears 3-4 chestnuts inside a very prickly shell. When ripe these will open up and the chestnuts can be carefully removed.

Where: South of the Alps whole chestnut forests exist. In the north the tree is rarer and the fruit does not always ripen sufficiently.

When: Autumn

Culinary uses: Chestnut is one of the most versatile wild foods, prepared both for nourishing sweet and savoury meals. The simplest way to enjoy chestnut is slitting the outer skin and then roasting them in a pan over a hot fire. Dried they can be ground to a flour to add to wheat flower to make bread or on its own for a pasta dough or chestnut pancakes. In Italy a sweet called Castagnaccio is very popular. Boiled and peeled it can be turned to a mash in place of mashed potatoes. A chestnut sorbet is delicious and of course there is the chestnut stuffing for a turkey or chicken. Together with white beans a sandwich spread can be made. Chestnut soup is also tasty.

Medicinal uses: Unlike most other nuts, chestnuts are low in fat, protein and cholesterol free but rich in complex carbohydrates. They are also gluten free. They are rich in mineral salts and a good source of vitamins C, B1 and B2 and folates. An infusion of the leaves is said to relieve whooping coughs.

Caution: Do not confuse the sweet chestnut with the horse chestnut (aesculus hippocastanum), which is great for playing conkers and making stick animals, but not for eating.

Recipe: Chestnut stuffing for chicken: Boil, peel & mince 450g of chestnuts. Fry 1 onion in some butter, add 450g sausage meat, 50g of sultanas, some sugar, cinnamon, a little tomato juice and the minced chestnuts, plus season to taste and cook until nicely browned. Leave to cool and stuff into your bird before roasting. Double quantities for a turkey.

Link:
http://www.pfaf.org/user/Plant.aspx?LatinName=Castanea+sativa
http://pathtoselfsufficiency.blogspot.com/2008/10/some-things-to-do-with-chestnuts.html

Chickweed (*stellaria media, stellaria pubera*)

Description: Chickweeds are an annual herb, widespread in temperate zones. They are as numerous in species as they are in region. Most are succulent and have white flowers, and all with practically the same edible and medicinal values. They all exhibit a very interesting trait, they sleep, termed the 'Sleep of Plants.' Every night the leaves fold over the tender buds and the new shoots.

Where: Common in gardens, fields and on disturbed ground

When: Spring to early Summer

Culinary uses: Whole plant harvested when just in flower in salads or briefly cooked like spinach and added to soups (see recipe under Sorrel). It is rich in nutritious value.

Medicinal uses: Relieves constipation, an infusion of the dried herb is used against coughs and sore throats, beneficial in the treatment of kidney disease. A decoction is used externally to treat rheumatic pains, wounds and ulcers and will help itching of the skin.

Recipe: Medicinal tea: To 1 tbls. dried herb, 2 if fresh, add 1 cup boiling water steep for 10 min. Take in ½ cup doses 2 to 4 times daily, during a cold or flu.

Link:
http://www.altnature.com/gallery/chickweed.htm

Chicory, Wild (*cichorium intybus*)

Description: The bright blue flowers of this common plant can be found along paths and on fields everywhere in Europe. The leaves resemble dandelion (see photo top right for comparison: left dandelion leaf, right chicory leaf) and in fact in the temperate zones of Europe I have yet to come across any dandelion look-alike that isn't edible, although some taste better than others.

Where: Along pathways and waste grounds.

When: Leaves for eating are best gathered in Spring before flowering, after that they become very bitter.

Culinary uses: Young leaves are best briefly blanched and drained before then sautéing them in a little oil, lemon juice and garlic as pot herb. A few young leaves and/or a few flowers added to a mixed salad add a bitter twang. A coffee substitute can be made from the dried root. The roots can also be used to season soups, sauces and gravies

Medicinal uses: A tea made from the leaves of chicory and/or dandelion together with plantain has a cleansing effect on the liver, relieves indigestion and acts as an appetizer.

Recipe: Fave e Cicoria: A recipe from Puglia in Southern Italy. Make a puree from 500g broad beans, by slowly boiling them in just a little water until a creamy texture is achieved. Add plenty olive oil & seasoning. Briefly blanch a generous handful of chicory leaves in boiling water, drain and sauté in some oil, lemon juice and garlic. Serve chicory on top of the pureed broad beans.

Link: http://www.pfaf.org/user/Plant.aspx?LatinName=Cichorium+intybus

Chives, Onion (*allium schoenoprasum*)

Description: At first glance young chive plants may be mistaken as common grass. In fact on my favourite location they grow on a field amongst grass. Closer inspection will reveal round hollow stems though with a distinct onion aroma when broken. Later in the summer purple clover-like flowers will appear.

Where: Rocky pastures and damp meadows, preferring calcareous soils

When: Spring - Summer

Culinary uses: Best used raw in salads, to flavour cream cheese, sprinkled on soups, added to an omelette, etc.

Medicinal uses: Chives stimulate the appetite, have a tonic effect on the kidneys and are said to help lower blood pressure.

Clary (*salvia sclarea*)

Description: A biennial member of the sage family it grows up to 60cm in height. The leaves are slightly crinkly and resemble those of peppermint. They exude an intense aroma. The flowers are purple and bell-shaped, typical of those of the sage family.

Where: Woodlands in the northern Mediterranean.

When: Spring and summer

Culinary uses: Fresh leaves used in fritters, home-made wine (giving it a muscatel-like flavour) and beer, omelettes, soups and salads in small quantities as it has strongly aromatic taste. It also makes a very agreeable tea.

Medicinal uses: It's name was originally coined by Culpeper as Clear-Eye, because the mucus rich seeds can be used to remove foreign bodies from the eye (see recipe). A compress is used externally to treat boils. It relieves indigestion and wind and is regarded as a calming tonic. Due to it's oestrogen stimulating action it is useful to treat menopausal complaints. It should however not be prescribed to pregnant women.

Recipe: Clary Eye Lotion: Soak about 6 clary seeds in clean, previously boiled, warm water. Leave to swell for a few minutes, then carefully introduce into the corner of the eye with a cotton bud. Particles of grid will adhere to the mucilage and make it easy to remove.

Link:
http://www.pfaf.org/user/Plant.aspx?LatinName=Salvia+sclarea

Clover, Red (*trifolium pratense*)

Description: The common red clover as we all know it, the shamrock is usually 3-leafed, or if you are really lucky 4-leafed. All varieties are edible, but the red flowering species is the best for both culinary as well as medicinal purposes.

Where: Meadows and grassy fields

When: Almost all year, but best harvested just as it comes into flower

Culinary uses: Young leaves and young flowering heads can be eaten raw or cooked. Leaves are best cooked like spinach. Dried and powdered they can be sprinkled over rice to impart a mild vanilla flavour. Seeds are best sprouted before being eaten like alfalfa. A delicate sweet tea is made from the fresh or dried flowers (see recipe).

Medicinal uses: High levels of phytoestrogens in the flowers help with menopausal issues. High concentrations of a decoction applied on tumours is said to encourage them to grow outwards. Internally the plant is used as treatment for skin complaints, cancers (especially of the breast and ovaries), chronic degenerative diseases, gout, whooping cough and dry cough.

Caution: When diseased the plant may contain toxic alkaloids, although this is currently being studied for it's anti-diabetic and anti-AIDS activity.

Recipe: Sweet Tea alla Mrs. H from Idaho: A health giving and pleasant tasting beverage made up from Echinacea, mint, clover, lemon balm and elder flowers sweetened with a bit of honey.

Link:
http://www.pfaf.org/user/Plant.aspx?LatinName=Trifolium+pratense

Comfrey (*symphtum officiale*)

Description: Up to 1.5 m tall plant with large, fleshy and hairy leaves. The mauve flowers droop in bell-like clusters at the tip of the plant during most of summer. When broken a mild cucumber aroma can be detected.

Where: Shady or semi-shady, moist places in woods, often near streams.

When: Best used when young in Spring to early Summer.

Culinary uses: Young leaves finely chopped can be cooked as spinach or coat in batter for a comfrey tempura. A tea can be made from the roots and dried leaves.

Medicinal uses: Used externally as a poultice the leaves of the plant have a long history of being used to treat cuts, bruises, sprains, sores, eczema, varicose veins and even broken bones. Internally the leaves and roots are used for the treatment of pulmonary complaints. An infusion of the leaves and roots is given to treat chest colds, to improve circulation and for the intestines.

Recipe: Comfrey Tempura: Make a batter with eggs, a little water, some rice flour and salt and pepper. Dip in comfrey leaves and deep fry a couple of minutes in hot oil until crisp. Serve with a soy sauce dip as an appetiser.

Caution: Many sources will warn against any use of this beneficial herb as it contains toxic alkaloids, which can have an accumulative effect and lead to liver damage. Many other sources claim that the benefits of this herb far outweigh any small risks. It is therefore not recommended for people with a history of liver problems and internally it should be taken in moderation.

Link:
http://www.herbsarespecial.com.au/free-herb-information/comfrey.html

Coltsfoot (*tussilago farfara*)

Description: The odd thing about this plant is that the flowers and leaves appear at different times and are rarely seen together at the same time. The yellow flowers appear in Spring, dandelion like heads, but on fleshy stems up to 25cm in height. When broken, the stems will exude a heady perfume. The leaves appearing later give the plant its name as they are about the size and shape of a(n un-shoed) horse's foot, although they can grow to the size of a dinner plate.

Where: Damp ditches, hedge banks, wastelands all over Europe

When: Early Spring for the flowers, late Spring for the young leaves.

Culinary uses: Flower buds and young leaves can be eaten raw or cooked as part of a wild salad or herbal soup to add a pleasant aniseed flavour. An aromatic tea can be made from the dried leaves and flowers. The dried and burnt leaves can be used as a salt alternative.

Medicinal uses: A tea made from the dried flower stems and leaves relieves coughs. It is best mixed with other herbs though such as thyme, aniseed, primrose and mullein. A poultice made from the flowers have a soothing effect on sores, insect bites and eczema. The dried leaves can be combined  with those of horehound and mullein as a tobacco alternative. It induces coughing and clears phlegm out of the lungs for those wanting to give up smoking and clear their lungs.

Caution: The plant contains liver damaging alkaloids. Do not use internally when pregnant, for children under the age of 6 or patients with known liver problems. Healthy people should consume only moderate quantities of this herb and not for prolonged periods of time.

Link:
http://www.pfaf.org/user/Plant.aspx?LatinName=Tussilago+farfara

Cornflower (*centaurea cyanus*)

Description: Once common in wheat fields but now in drastic decline due to modern agricultural practices. The bright blue flowers stand out a mile from the golden wheat.

Where: In wheat fields and on porous, nutrient rich soils.

When: Summer

Culinary uses: Fresh flowers can be used in salads. Young shoots and flowers can also be cooked as a vegetable.

Medicinal uses: An infusion can be used as a treatment for constipation, as a stimulant, to improve digestion and supporting the liver or as a mouthwash for ulcers and bleeding gums.

Link:
http://www.pfaf.org/user/Plant.aspx?LatinName=Centaurea+cyanus

Corn Salad (*valerianella locusta*)

Description: The common corn salad or Lewiston cornsalad can also occasionally be found in the wild. From a rosette of spatula-shaped light green leaves rises a stem of up to 30cm which bears small posies of white flowers.

Where: Cultivated grounds and hedge banks all over Europe. It prefers some dappled shade and dry soils.

When: Early spring to summer

Culinary uses: The leaves make on of the best salad ingredients especially with a light dressing of onion and garlic.

Medicinal uses: None known

Link:
http://www.pfaf.org/user/Plant.aspx?LatinName=Valerianella+locusta

Costmary (*tanacetum balsamita*)

Description: Also known as alecost due to the fact that it used to be used to flavour beers instead of hops, it is a perennial plant growing up to 1m tall with large, oval, lightly serrated leaves and small, button-like green flowers turning yellow later. When rubbed all parts of the plant give off what has been described as a delicate balsamic aroma to a strong smell of bubblegum. It is a native of Eastern Asia but has been naturalised in many areas of Europe including Britain, where it once was a common garden herb.

Where: It has become quite an invasive weed on cultivated grounds.

When: Spring to summer

Culinary uses: Moderate amounts of the leaves, raw or cooked, add a sharp tang to salads, soups, game stuffings and fruit cakes. Also can be used as part of an omelette or a flavouring in ales. The leaves make a pleasant tea.

Medicinal uses: Not much used in modern herbalism, but the tea is said to relieve colds, catarrh, upset stomachs and cramps. Externally crushed leaves are applied to insect stings and burns to relieve itching. It also makes a pleasant addition to a potpourrie.

Links:
http://www.botanical.com/botanical/mgmh/c/costm107.html

Cowslip (*primula veris*)

Description: Very similar in appearance to the more common primrose, but the pale yellow flowers stand a little taller and are more bell-shaped. Leaves are oval and fleshy. This plant has now become quite rare and is protected in some countries.

Where: It prefers more open locations then the common primrose such as meadows and fields.

When: Spring

Culinary uses: Young leaves can be added to soups. Dried leaves make a pleasant tea. Flowers make a decorative addition to a salad. A country wine can be made from the flowers if it is found in abundance.

Medicinal uses: The whole plant is used in the treatment of spasms, cramps and rheumatism. It contains salicylates, which are one of the active ingredients of aspirin and contributes to its anti-inflammatory effects. The flower is anti-spasmodic and has a sedative effect on hyper-active children. The plant is also know to be effective in the treatment of bronchitis.

Caution: Some people may be allergic to this plant, although allergic reactions are easily treated. Extensive and prolonged use of this plant may interfere with high blood pressure treatments.

Link:
http://www.botanical.com/botanical/mgmh/c/cowsl112.html

Cuckoo Flower (*cardamine pretensis*)

Description: Also known as Lady's Smock, this perennial flowering plant grows to about 50cm in height. The narrow leaves are pointy and the 1-2cm flowers have 4 pale pink petals.

Where: Moist and slightly shady places on meadows and along streams and edges of ponds. As a wetland plant it is becoming rare in some places now.

When: Spring

Culinary uses: The young leaves have a pungent, spicy flavour reminiscent of watercress. The leaves and flowers can be added to salads. Just the leaves are also cooked as a potherb.

Medicinal uses: The plant is rich in vitamin C and iron. A tea made from 2 teaspoons of chopped herb per cup of boiling water taken a couple of times a day makes a good spring cleanser.

Link:
http://www.pfaf.org/user/Plant.aspx?LatinName=Cardamine+pratensis

Curry Plant (*helichrysum italicum*)

Description: On hot summer days these fragrant plants can be smelled before they can be seen. At the leaf stage they somewhat resemble lavender, but the flowers are yellow knobs and the smell is distinctly curry-like.

Where: On sunny spots on rocky or sandy soil around the Mediterranean.

When: An evergreen, the leaves can be picked all year around, but best in summer.

41

Culinary uses: Leaves can be sparingly sprinkled over salads although the curry smell does not transfer well onto the flavour and is usually disappointing. A few finely chopped leaves sprinkled onto soups as a garnish works too

Medicinal uses: An oil made from the flowers is said to be anti-inflammatory, anti-fungal and soothing to raw chapped skin and burns.

Link:
http://pfaf.org/user/Plant.aspx?LatinName=Helichrysum+italicum

Daisy (*bellis perennis*)

Description: This well known and loved, common low-growing plant, flowering throughout much of the year with it's ground-hugging small oval leaves is actually edible.

Where: Lawns, fields and meadows

When: Almost all year around in mild climates, but best gathered in spring

Culinary uses: The young leaves and flowers make a pretty addition to mixed salads. The leaves can be cooked like spinach or sautéed with a little lemon and garlic.

Medicinal uses: A tea made from the leaves and flower heads helps against colds and rheumatism and calms upset stomachs. It is also a blood purifier and can be taken as a spring cleanser for the liver and kidneys. Chewing fresh leaves relieves mouth ulcers. Externally the tea can be applied to eczema and small open wounds reluctant to heal.

Link:
http://www.pfaf.org/user/Plant.aspx?LatinName=Bellis+perennis

Dandelion (*taraxacum officinalis*)

Description: Another well known and easily found herb with it's distinct lion toothed leaves and bright yellow flowers. All parts exude a milky substance when broken. There are many dandelion look-alikes, none of them poisonous as such, but some more pleasant tasting then others.

Where: Meadows, lawns, fields and cultivated areas everywhere

When: Both leaves and flowers best collected in spring

Culinary uses: Young leaves can be used raw in salads or cooked as a pot herb. The inherent bitterness can be lessened by excluding light from the growing leaves for a couple of weeks before harvest, however much of the nutrients are also lost during this process. Briefly boiling and then changing the water has a similar effect. Best though to make a mixed salad with other less bitter ingredients. The flowers are also edible and make a pretty addition to a salad or can be made into dandelion wine. Unopened flower buds can be preserved in vinegar and used like capers. Dandelion is extremely nutritious containing protein, carbohydrates, calcium and various minerals and vitamins. The dried and roasted roots of two year old plants make a coffee substitute.

Medicinal uses: The root is the most effective part of the plant, but all parts are beneficial. A tea is used in the treatment of gallstones, urinary disorders, liver problems, dyspepsia with constipation, high blood pressure, gout and eczemas amongst other things. It's strongly diuretic without depleting the body of potassium. The plant is also known to be anti-bacterial. The latex in the plants sap can be used to remove corns and warts.

Link:
http://www.herbsarespecial.com.au/free-herb-information/dandelion.html

Dock (*rumex*)

Description: Of the many different species of the rumex family the best for eating is sorrel, which is described elsewhere in this book. All the other species are edible, but few are really tasty as they all are rather bitter. However young leaves of this family can be used in moderate quantities. The most common examples are the Fiddle Dock (r. pulcher), named after its fiddle shaped leaves, Broad-Leaved Dock (r. obtusifolius), pictured above and Curled Dock (r. crispus) which is self-descriptive. They all have a fairly rough leave structure in common and produce inconspicuous flowers on long stems.

Where: All are very common on waste grounds

When: Best in late winter to early spring

Culinary uses: Young leaves can be cooked and mixed with other greens as a vegetable. It is best to change the water at least once during cooking to remove some of the bitterness. Cooked they can be added to soups and vegetable tarts. One of the nicest ways to use them is stuffed in place of vine leaves (see recipe). Very young fiddle dock leaves can be added raw to salads adding a sharp tang a bit like sorrel.

Medicinal uses: Leaves of dock are applied externally to blisters and burns and are a common folk remedy against nettle stings. They also contain vitamin C. Curled dock makes a safe laxative.

Recipe: Stuffed Dock Leaves: Pour boiling water over a dozen or so young dock leaves and leave for a few minutes before draining. Fry a chopped onion and some garlic in olive oil until soft. Stir in a handful of rice, some tomato paste and a glass of water and simmer until half the liquid is gone and the rice half cooked. Add some fresh mint, dill, a pinch of cinnamon, some pine kernels, raisins and salt and pepper. Place stuffing onto prepared leaves and make little parcels. Line a frying pan with some broken dock leaves, add oil, lemon juice and garlic and place the leaf parcels on top, ends down. Pour over some more water and simmer covered for 1 hour, checking liquid now and again, until soft. Serve with a squeeze of lemon juice as an antipasto

Caution: All species of the rumex family contain oxalic acid, which makes certain minerals unavailable to the body, which in turn can lead to deficiencies. Consume in moderation as part of a varied diet.

Link:
http://www.botanical.com/botanical/mgmh/d/docks-15.html

Echinacea (*echinacea purpurea*)

Description: A native to north America, this pretty flower from the asteraceae family can sometimes be found naturalised in Southern Europe. It grows up to over 1 metre tall with 8 cm wide pinkish purple flower heads with slightly bent back petals.

Where: Open fields and cultivated ground

When: Summer

Culinary uses: As part of a tea (see recipe under clover)

Medicinal uses: One of the most effective de-toxicants for the circulatory, lymphatic and respiratory systems. The plant has a stimulant effect on the immune system helping prevent flu attacks. Externally the root is used in the treatment of sores, wounds and burns with its anti-bacterial action.

Link:
http://www.pfaf.org/user/Plant.aspx?LatinName=Echinacea+purpurea

Elder (*sambucus nigra*)

Description: Up to 4m tall deciduous shrub with up to 10cm long oval leaves, displaying umbrella like, very fragrant white flowers and later clusters of deep black berries

Where: Hedgerows everywhere

When: Flowers in spring, berries late summer, early autumn

Culinary uses: Both flowers and berries make excellent country wines. Flowers as fritters (see recipe), turned into a refreshing cordial, an ingredient in cakes or added to gooseberry jam. Dried flowers can be made into tea. Berries can be cooked into jams or chutneys or as a topping on pies. If frozen first, the berries are easier to strip off the stalks.

Recipe: Elderflower Fritters: Carefully cut the flowers off the shrub with scissors, leaving a fair bit of the stem. Dip the flower heads in a slightly sweetened light pancake batter and fry in some butter. While one side is frying, cut off the stalks from the other. Turn, frying the other side until brown. A delicious spring time treat.

Elderberry Chutney: Simmer 500g of onions in ½ litre vinegar until soft. Add 2kg elderberries (stripped off their stalks), 200g raisins, salt, chilli, mustard powder and pickling spice (recipe see Juniper) and boil until soft. Add another ½ litre vinegar and 250g of brown sugar. Simmer until it thickens. Bottle in hot sterile jars.

Medicinal uses: The fruit is high in vitamin C and Potassium is helpful against colds and strengthens the immune system. They are also rich in anthocyanins a powerful antioxidant. The flowers are used externally, mostly in the form of 'elderflower water', to relieve burns and inflammations.

Caution: Stalks, leaves and under-ripe berries are toxic and should be avoided, although they can be used in herbal medicine for their purgative qualities. Even ripe berries are best cooked to remove any toxins contained in them.

Link:
http://elderberries.ning.com/

Fennel (*foeniculum vulgare*)

Description: A perennial plant growing up to 2m, with fine feathery foliage and yellow umbrella-like flowers. Rubbed it gives off a strong anise-like smell.

Where: Native to the Mediterranean, where it grows along many dry banks in sunny positions.

When: Foliage spring to autumn; seeds in autumn.

Culinary uses: The leaves can be finely chopped and used in salads, soups, stews, sauces and with fish. A tea can be made from the leaves and seeds. The seeds are great with a pork roast together with chilli flakes and rosemary. The swollen bulbs of Florence fennel can be eaten raw or cooked as a vegetable. A tasty tea is made from leaves and/or seeds.

Medicinal uses: All parts of the plant have a carminative effect, especially the seeds and are generally beneficial to the digestive system. A tea made by pouring boiling water over a teaspoon of seeds is used to combat flatulence and stomach pains. An infusion of the root is helpful to combat urinary disorders. It is also said that fennel improves eyesight. Fennel seeds are anti-bacterial.

Recipe: Wild fennel and chickpea soup: Sauté an onion in olive oil. Add some garlic and a bunch of finely chopped fennel leaves. Add a can of chickpeas and some water. Bring to the boil and cook for 15 minutes or so.

Season with salt and pepper and add some vermicelli soup noodles. Bring back to the boil and cook until pasta is done.

Fennel, orange and black olive salad: Combine roughly chopped Florence fennel bulb, pared and roughly chopped oranges, stoned black olives and roughly shaven parmesan cheese and drizzle all with good olive oil to make a delicious winter salad.

Caution: Some people may be allergic to the sap or essential oil of the plant, causing hypersensitivity to sunlight or dermatitis on contact with the skin.

Link:
http://www.pfaf.org/user/Plant.aspx?LatinName=Foeniculum+vulgare

Feverfew (*tanacetum parthenium*)

Description: Small bushy perennial plant up to 50cm high with distinctly light green, fine foliage and lots of small daisy-like flowers.

Where: Wastelands, walls and along pathways.

When: spring to autumn

Culinary uses: Not really

Medicinal uses: Chewing a leaf or two every day is one of the best preventatives for migraines, as tested by myself. An infusion made from the leaves and flowers relieves menstrual pains and lowers fevers. It is also used in the treatment of arthritis and rheumatism. Use in footbaths to relieve swollen feet.

Caution: Should not be ingested during pregnancy. May cause mouth ulcers with some people.

Link:
http://www.altnature.com/gallery/feverfew.htm

Fig (*ficus caricacarica*)

Description: Large deciduous tree. Leaves up to 25x18cm with 3 or 5 deep lobes. Fruit green, sometimes purple skinned up to 10cm long oval shapes. When cut open, ripe fruit displays pink to deep red flesh. When broken the stalk emits a milky substance, which can be an irritant to the skin

Where: On sunny or semi-shaded spots all around the Mediterranean. It can even grow out of sheer rock faces or cracks in walls.

When: Most trees have two fruiting seasons, one in early summer and one in late summer. The first fruit tends to be smaller, but sweeter.

Culinary uses: I am always amazed how many people only know the dried version of this delicious fruit. Eat it, raw as it is, simply suck out the sticky insides. If you have lots make jam, fig sauce for meat dishes, fig cake or of course dry some for later use

Medicinal uses: Figs are one of the highest plant sources for calcium and fibre. They are also rich in copper, manganese, magnesium, calcium, and vitamin K making them one of the most nutritious fruits around. They are known to have a laxative effect and contain many anti-oxidants. Dried figs have all these in greater concentration of course. The latex from the stems is used to treat corns, warts and piles. The roasted fruit can be used as a poultice on dental abscesses and gum boils. The plant has anti-cancer properties.

Recipe: Fig Cake: Beat 100g soft butter, 150g sugar and vanilla essence until creamy. Beat in 2 eggs, then 70g ground hazelnuts and finally zest and juice of 1 lemon. Puree 10 fresh figs and add to the mixture and finally combine with 250g wholemeal flour and some baking powder. Pour into a greased round baking tin, decorate the top

with a few more fresh figs cut open and bake at 175C for 40 minutes. A lovely moist fruit cake!

Figs and Goat Cheese: Open up the fresh fruit and drop a dollop of fresh goat's cheese on top and sprinkle with some pepper. Bake a few minutes in a hot oven until cheese is melted and serve as a delicious starter.

Caution: The sap can be an irritant to skin and eyes.

Link:
http://www.pfaf.org/user/Plant.aspx?LatinName=Ficus+carica

Fool's Watercress (*apium nodiflorum*)

Description: Up to 60cm tall, semi-aquatic celery-like plant with pairs of about 8cm long leaves and hollow, segmented stems. All parts of the plant exude a distinct fresh aroma of sweet celery with a hint of cinnamon. The flowers consist of small white umbels.

Where: At the edges of streams and waterways and in damp ditches.

When: spring to summer

Culinary uses: The leaves add a very pleasant flavour to a mixed salad. The stems and leaves together can be cooked and added to a soup in the same way as celery.

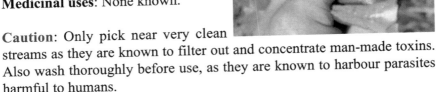

Medicinal uses: None known.

Caution: Only pick near very clean streams as they are known to filter out and concentrate man-made toxins. Also wash thoroughly before use, as they are known to harbour parasites harmful to humans.

Link: http://en.wikipedia.org/wiki/Fools_watercress

French Scorzanera (*reichardia picroides*)

Description: A dandelion look alike also known as Common Bright Eyes. From a centre of toothed shiny leaves emerges an up to 30cm tall flowering stem with dandelion-like flowers. The roots are distinctly swollen like small white carrots. It is easily confused with other plants of the same genus, but the taste test reveals a much sweeter flavour than most of its cousins.

Where: On fields and in semi-shaded areas along hedges.

When: Spring to summer.

Culinary uses: Leaves, both young and mature, make a very agreeable lettuce substitute. They can also be cooked as a potherb or lightly browned in olive oil with some garlic. The roots can also be eaten either raw or cooked.

Medicinal uses: Non known

Link:
http://www.pfaf.org/user/Plant.aspx?LatinName=Reichardia+picroides

Garlic, Wild *(allium)*

Description: Ramsons, a.ursinum, have long, thin pointy leaves, similar to those of lily of the valley (which is poisonous!) and exude a strong garlic smell which can be detected from quite some distance. Cultivated Garlic (a. sativum) has also naturalised in much of the Mediterranean. Other common varieties are Three-Cornered Leek, (a. triquetrum, photo right and below), with its distinct triangular stalks and Rosy Garlic (a. roseum, photo bottom), displaying little leaf, but pretty posies of rosy-white flowers.

Where: Ramsons can be found in moist woodlands all over central Europe. A. sativum is found on dry, open and sunny locations around the Mediterranean. A. triquetrum is found in moist locations around the Mediterranean. A. roseum is common on meadows.

When: Ramsons, being mostly used for their leaves, are best picked in early spring whilst true garlic is best picked in high summer. Ramson flowers and bulbs can also be eaten. Bulbs are best harvested between early summer to early winter. A. triquetrum and a. roseum are best picked in spring for their foliage and flowers.

Culinary uses: In the case of ransoms and a. triquetrum the leaves are mostly used to add chopped to salads or as a pesto as with basil. It also makes a delicious soup. The flowers and bulbs have a stronger flavour than the leaves. Common garlic has too many uses to even start listing here. The flowers of rose garlic make a delicious salad ingredient.

Medicinal uses: All types of garlic have similar health properties, although sativum is most effective. Garlic is known as a powerful natural antiseptic. It contains vitamins A,B and C as well as copper, sulphur, manganese, iron and calcium. It cleanses the intestines, helps lower blood pressure, expels worms, helps ward off colds and alleviates rheumatism. It is beneficial for  elderly people suffering from hardening of the arteries and general ageing symptoms and it is said to help people suffering from sinus trouble and hay fever. It is also said to improve the skin, help diabetes sufferers and help prevent heart attacks. In short, it is one of the most useful herbs around.

Caution: Large quantities have been known to be mildly poisonous to some mammals, especially dogs. Take care not to confuse ramsons for lily of the valley. The smell is a definite identifier. If it doesn't smell of garlic, it's not garlic.

Link:
http://www.pfaf.org/user/Plant.aspx?LatinName=Allium+sativum

Golden Fleece, Smooth (*urospermum dalechampii*)

Description: A perennial relation of the dandelion, it looks superficially similar, but with more solid, sulphur-yellow flowers, not dissimilar to carnations. It grows to a height of up to 50cm on a usually un-branched hairy stem. The taste is somewhat more bitter than dandelion.

Where: On meadows and along pathways along the Mediterranean.

When: Spring for the young leaves

Culinary uses: Young leaves raw added to salads with other sweeter ingredients or boiled then sautéed with tomato and garlic.

Medicinal uses: None known

Link:
http://anentangledbank.wordpress.com/2012/05/13/urospermum-dalechampii/

Goldenrod (*solidago virgaurea or solidago canadensis*)

Description: Up to 1.2 metre tall upright plant with off-set pointy leaves and bright yellow flowers growing in large groups together. When rubbed the leaves give off a strong aromatic scent reminiscent of chamomile. Both variants, s. virgaurea and s. canadensis, offer similar properties, the former being native to Europe and the latter native to North America.

Where: The edge of woodlands, meadows and along pathways.

When: Late summer into autumn

Culinary uses: Not used

Medicinal uses: A tea made from the leaves and flowers is one of the most effective cures against infections of the urinary tract and kidney problems. It also helps with digestive disorders. Externally it is used on wounds and sores to speed up the healing process. Sufferers of influenza and whooping cough are also said to find relief from this plant.

Link: http://www.herbs2000.com/herbs/herbs_goldenrod.htm

Good King Henry (*chenopodium bonus-henricus*)

Description: A perennial, leafy herb growing up to 30cm in height. The leaves are 5-10 cm long, diamond or triangular shaped and have a slightly waxy, succulent texture.

Where: Rich pastures and roadsides all over Europe

When: Spring to early summer

Culinary uses: The leaves cooked make a good spinach substitute. The young shoots in spring can be eaten like asparagus.

Medicinal uses: A rich source of iron and vitamins. A tea can be made to help patients suffering from kidney problems or rheumatism. The seed has a gentle laxative effect.

Caution: The plant does contain small quantities of saponins (the same toxin contained in many beans) and oxalic acid, both slightly toxic. However most will be broken down when cooking the plant and the rest will only be poorly absorbed when passing through the digestive system, rendering them harmless. However people with a tendency towards gout, kidney stones or arthritis should exercise caution when using this herb.

Link:
http://www.pfaf.org/user/Plant.aspx?LatinName=Chenopodium+bonus-henricus

Goosegrass (*galium aparine*)

Description: Long (up to 2 metres) straggling weed with groups of 6 elliptical leaves around the stem, small white flowers and tiny hooks which attach themselves to clothing. Also known as cleavers.

Where: As a weed on cultivated ground, meadows and hedgerows

When: Spring to summer

Culinary uses: Tender young shoots are cooked as a pot-herb, best mixed with other wild greens. Add a handful to a vegetable soup. The small hooks will dissolve when cooked. The roasted seed makes a coffee substitute and the whole plant dried can be used as a tea substitute.

Medicinal uses: it is a cleansing, diuretic tonic that stimulates the lymphatic system, helping eczema, psoriasis, arthritis and liver disease.

Caution: The sap of the plant can cause dermatitis in sensitive people

Recipe: Boil briefly together with some young nettle shoots and serve as a side green with a knob of butter and well seasoned. To make coffee substitute, pick fruits in June and July, roast for an hour at 150 C until dark brown. Grind the result in a pestle and mortar, boil a handful in a litre of water until coffee coloured and strain.

Link:
http://www.pfaf.org/user/Plant.aspx?LatinName=Galium+aparine

Gorse *(ulex europeus)*

Description: Evergreen shrub up to 2m tall covered in vicious thorns and may be seen to show off its bright yellow flowers at any time of the year.

Where: Open sunny locations on poor soils all over Europe.

When: All year

Culinary uses: Flowers make one of the nicest country wines. They can also be made into a tea with a distinct coconut-like flavour or sprinkled into a salad.

Medicinal uses: Gorse plays no significant role in herbalism.

Link:
http://www.pfaf.org/user/Plant.aspx?LatinName=Ulex+europaeus

Ground Elder (*aegopodium podagraria*)

Description: A rapidly spreading weed on disturbed ground where it spreads via rhizomes and quickly covers a large area smothering all other vegetation in its path. It gets its name from the superficial resemblance to the elder shrub of both the leaves and the flowers. However it only grows up to 1m tall. It is also known as bishops weed or gout weed.

Where: On disturbed ground all over Europe except Iberia

When: Leaves best picked in spring.

Culinary uses: According to some it makes a delicious spring vegetable cooked as a potherb, others dislike its tangy, slightly medicinal flavour.

Medicinal uses: One of its common names describes the plants usefulness in preventing and treating gout. The juice of the leaves applied to insect

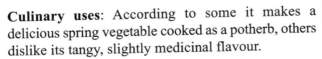

56

bites and small cuts relieve itching and promote healing. The herb is also used as a poultice in the treatment of rheumatism and painful joints.

Link:
http://www.pfaf.org/user/Plant.aspx?LatinName=Aegopodium+podagraria

Ground Ivy (*glechoma hederacea*)

Description: A perennial evergreen creeping plant from the mint family. It displays fan-shaped delicate leaves up to about 4cm in diameter and have a smooth, shiny surface. The flowers are bluish to purple and funnel shaped. Leaves rubbed give off an aromatic, slightly unpleasant smell

Where: Hedgerows and woodland edges as a groundcover in dappled shade.

When: Spring.

Culinary uses: Young leaves add an aromatic tang to salads. It can be mixed with cream cheese or crème fraîche on new potatoes. The leaves can also be cooked as a potherb. A herbal tea made from the leaves is rich in vitamin C. It used to be added to beer as a preservative and flavouring before hops.

Medicinal uses: The tea has a beneficial effect on stomach and intestinal problems. Externally an infusion is applied to heal wounds, as a gargle against throat infections and as a rinse to treat eye infections. The herb is also beneficial for the kidneys. Leaves for medicinal purposes are best gathered in May and dried for later use.

Caution: There are some reports of toxicity. Avoid if pregnant and consume in moderation.

Link:
http://www.pfaf.org/user/Plant.aspx?LatinName=Glechoma+hederacea

Hawkbit, Tuberous *(leontodon tuberosus)*

Description: A dandelion look-alike, with less indented and slightly hairy leaves. The flowers are also similar to those of dandelion. The roots are the definitive identifier for this species as they swell into a series of small tubers, which, like the leaf, have a pleasant non-bitter taste.

Where: Meadows, cultivated ground and along pathways in Southern Europe.

When: Autumn to spring

Culinary uses: Both the leaves, young and mature, and the roots make an excellent salad ingredient.

Medicinal uses: None known

Link:
http://www.piantespontaneeincucina.info/documenti/schede_delle_principa li_specie_della_tradizione_alimentare/leontodon_tuberosus.pdf (Italian)

Hawk's Beard *(crepis leontodontoides)*

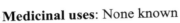

Description: A dandelion look-alike, with smaller more delicate leaves. Also the flowers are similar to those of dandelion. Known as radichella in Italian. There are various other species of the crepis genus including Italian Hawks Beard (c. bursifolia), c. sancta, Beaked Hawks Beard (c. vesicaria), False Hawks Beard (c. japonica), which can be used in similar ways.

Where: Meadows, cultivated ground and along pathways as well as crevices in walls.

When: Spring

Culinary uses: In mixed salads and cooked as a potherb with other non-bitter greens.

Medicinal uses: It is diuretic and assists digestion.

Link: http://www.eattheweeds.com/crepis-japonica-seasonal-potherb-2/

Hawthorn (*crataegus monogyna*)

Description: A dense deciduous tree or shrub growing up to 10m tall with dark smooth bark, small sharp thorns and small lobed leaves. In spring it is covered in dense white flower clusters that turn to red berries in the autumn containing one or two seeds each.

Where: Hedgerows and woodlands.

When: Flowers and leaves in spring, berries in autumn.

Culinary uses: Fresh young leaves can be added to a salad. Berries can be eaten raw, but are not very tasty, so are best used in hedgerow jams or jellies or as haw ketchup (see recipe). Dried and ground berries can be used as flour substitute to make bread. About 20% added to wheat flour makes a tasty mixture. The seeds roasted and ground make a good coffee substitute. A tea can be made from leaves, flowers and berries. The dried leaves can be smoked as a tobacco substitute.

Medicinal uses: Hawthorn is one of the most important herbal remedies, especially for heart complaints. Leaves, flowers and berries all have beneficial effects on the heart, keeping arteries elastic and blood flow smooth by inhibiting angiotensin converting enzymes responsible for restricting blood vessels. This also helps lessen pain in the heart area and increases warmth in cold hands and feet. Hawthorn is also rich in antioxidants and Vitamin C, slowing and even reversing signs of ageing and reducing inflammations and it contributes to relieving congestion, stagnation, sluggishness. Triperpene acids help balance low blood pressure and can help lower cholesterol. It is normally used as a tincture (see recipe below) or tea.

Recipe: Haw ketchup: Gently cook 750g of cleaned haws together with 450ml red wine vinegar for 30 minutes. Add some balsamic vinegar for flavour. Press mixture through a sieve, return to pan with 100g brown sugar, 2 tbsp salt and some pepper and simmer for another 10 minutes, adding more vinegar if necessary. Bottle in clean jars and use as you would tomato ketchup. Haw tincture: Use 225g of dried or 310g of fresh leaves and flowers and/or berries to 1 litre of liquid. Mix this in a food processor with a bottle of high strength vodka (min. 45%) or potable alcohol. Pour into an airtight jar and leave in a cool dark place. After 2 days add an extra 20% water if using 45% vodka, 50-60 % if higher percentage alcohol is used. Leave for another 2-4 weeks, then strain through a jelly bag and bottle in dark bottles. Take 1 tsp diluted with 5 tsp of water 2-3 times a day.

Link:
http://www.pfaf.org/user/Plant.aspx?LatinName=Crataegus+monogyna

Hazel (*corylus avallana*)

Description: Large deciduous shrub or small tree with simple serrated leaves, flowers in early spring in the shape of catkins and yielding the common hazelnut or cobnut in autumn, which is encased in a kind of cup.

Where: Hedgerows and woodlands all over Europe.

When: Nuts in autumn.

Culinary uses: Can be eaten raw or roasted. Mixed with herbs, breadcrumbs and egg they are shaped into nut burgers or nut roast. They make a good base for biscuits and cakes or can be roasted and chopped to top a salad. Hazelnuts are very nutritious and rich in proteins.

Medicinal uses: Not much used in herbalism. The finely ground nuts can be used as an ingredient in a facemask.

Recipe: Hazelnut butter: Crack hazelnuts and roast in the oven for some 20 minutes until lightly browned and exuding a pleasant aroma. In a food processor combine roasted nuts with some butter, a pinch of salt and a bunch of parsley. Use as a nutritious and satisfying sandwich filling instead of the ubiquitous peanut butter.

Link:
http://www.pfaf.org/user/Plant.aspx?LatinName=Corylus+avellana

Heather (*calluna vulgaris*)

Description: A low ground covering shrub with small hard leaves and pink to purple flowers

Where: On acid soils on moors or high plateaus

When: Late summer for the flowers

Culinary uses: The dried flowers are used as a tea or a flavouring for ale instead of hops.

Medicinal uses: Tea helps alleviate kidney and bladder problems. Flavoured with honey it's used as a night time drink to induce sleep. It has also been used in the treatment of rheumatism, gout and arthritis.

Recipe: Moorland tea: Robert Burns is supposed to have drunk a tea made from heather tops, leaves of bilberry, blackberry, speedwell, thyme, and wild strawberry.

Link:
http://www.pfaf.org/user/Plant.aspx?LatinName=Calluna+vulgaris

Herb Bennet (*geum urbanum*)

Description: Also known as Wood Avens, Colewort, Old Man's Whiskers, this is a delicate perennial flowering plant growing up to 50cm tall on thin wiry stalks topped by bright yellow, five-petalled flowers throughout the summer months.

Where: In woodland clearings or along hedgerows in semi-shaded, moist locations.

When: Autumn for the roots.

Culinary uses: The roots have a distinct clove-like flavour and can be used in the same way. The root is best used fresh as it looses much of its flavour when dried. Young leaves can be cooked as a potherb.

Medicinal uses: An infusion with the chopped up dried herb can be used as a mouthwash against gum infections and mouth ulcers. All parts of the plant are anti-inflammatory and antiseptic. An infusion used internally helps relieve upset stomachs and diarrhoea. Externally as a wash it is used in the treatment of haemorrhoids and vaginal discharges as well as various skin disorders.

Link:
http://www.botanical.com/botanical/mgmh/a/avens083.html

Hogweed, Common (*heracleum sphondylium*)

Description: This perennial plant, also known as Cow Parsnip, is a member of the umbelliferous family. It grows to a height of about 1.5m, which is its main distinguishing feature from Giant Hogweed (heracleum mantegazzianum), which grows to a height of up to 5m and is highly toxic even to touch (see caution below). From the root a ridged, hairy stem emerges, which produces up to 50cm long, hairy and serrated leaves with 3-5 lobes. The umbellate flowers have a slight pink tinge to them.

Where: On rich soils along pathways, especially in mountainous regions and on woodland edges.

When: Spring for the young stalks.

Culinary uses: The young stems, leaves and shoots can be eaten cooked as a potherb. The initial unpleasant aroma gives way to fruity smells of mandarin and carrot if cooked in saltwater for some minutes. The seeds and roots can also be consumed.

Medicinal uses: Roots and leaves are considered an aphrodisiac. They also aid digestion and have a sedative effect. However they are not used much now except for in homeopathy.

Caution: The fresh juice of the similar Giant Hogsweed, but to a lesser extend also that of this species is known to cause dermatitis on contact with the skin, leading to permanent scarring, photosensitivity and even cancer. Contact with the eyes can lead to blindness. Great caution should therefore be exercised. Avoid if in doubt or you have known sensitivity to furanocoumarins, the active toxin often found in this plant.

Link:
http://www.pfaf.org/user/Plant.aspx?LatinName=Heracleum+sphondylium

Hops (*humulus lupulus*)

Description: Tall climbing plant which winds itself around other plants for support. When young the shoots look very much like thin green asparagus although the stems are more square and slightly bristly. Leaves are toothed and 3 to 5-lobed. Female flowers appear in late summer in the shape 3cm green, papery cones.

Where: Hedgerows, wood clearings, alongside waterways

When: Young shoots in early spring, flowers in late summer.

Culinary uses: Hop flowers are best known for their use as flavouring and preservative in beer. Steeped in alcohol they can also be made into a liqueur or together with the leaves brewed into a tea. The young shoots in early spring are used and eaten like asparagus, in fact they are almost indistinguishable from wild asparagus and taste very similar.

Medicinal uses: Tea is a nerve tonic, a mild sedative and a muscle relaxant. It contains oestrogen, which increases lactation in nursing mothers. It also stimulates digestion and increases gastric flow. Hops are widely used as a folk remedy to treat a wide range of complaints, including boils, bruises, calculus, cancer, cramps, cough, cystitis, debility, delirium, diarrhoea, dyspepsia, fever, fits, hysteria, inflammation, insomnia, jaundice, nerves, neuralgia, rheumatism, and worms.

Caution: Can cause skin allergies in sensitive people.

Recipe: Hop Frittata: Fry a bunch of chopped hop shoots and a chopped onion in some olive oil for a few minutes. Beat 4 eggs with a tbsp of breadcrumbs and some grated hard cheese and add. Fry each side for a few minutes until set and lightly browned.

Link:
http://www.pfaf.org/user/Plant.aspx?LatinName=Humulus+lupulus

Horsetail (*equisetum arvense*)

Description: Horsetail is an ancient and strange plant. It has separate sterile non-reproductive and fertile spore-bearing stems, growing from a perennial underground rhizoamatous stem system. The fertile stems produced in early spring are non photosynthetic. They look rather like blanched asparagus spears or to put it more crudely, phallic. The sterile stems are up to 1 m tall and look like nothing else in the plant world, most reminiscent to perhaps a pine tree, but with hollow, segmented stems.

Where: Waste grounds, moist open ground, arable fields all over northern and central Europe and the northern part of Southern Europe, as far as Central Italy.

When: Spring for the edible shoots, summer for the sterile greens.

Culinary uses: Used with extreme caution (see below)! The fertile shoots in spring can be eaten as asparagus substitute, changing the water 3-4 times while cooking. It is considered a delicacy in Japan. Other parts are edible, but not worth the effort.

Medicinal uses: Horsetail is rich in silica and several alkaloids and minerals. It makes an excellent blood clotting agent, stemming bleeding. The green, non-fertile parts of the plant are gathered in summer and dried to use externally in hot baths to treat rheumatism. It is also a potent heart and nerve sedative, but caution should be taken not to overdose.

Caution: The plant contains the enzyme thiaminase, which in large doses is highly toxic as it inhibits the bodies intake of the Vitamin B complex. The enzyme is destroyed through heat or drying though and people with a healthy, well-balanced diet should suffer no negative consequences from moderate consumption.

Link:
http://www.pfaf.org/user/Plant.aspx?LatinName=Equisetum+arvense

Hyacinth, Tassel or Grape (*muscari comosum or m. botryoides*)

Description: The bulbs of either the Tassel or Grape Hyacinth, which is mostly known by its Italian name of lampascione, is considered a delicacy in southern Italy. It is a small flowering bulb, the flower appearing in summer in the shape of loose purple clusters. The bulbs are up to 3cm in diameter in case of the larger tassel hyacinth and about 1.5cm for the grape hyacinth.

Where: On dry fields, woodland edges and hedgerows all over southern Europe.

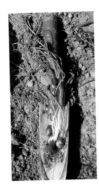

When: The bulbs are best dug up in spring as soon as the green shoots show. In fact I often accidentally dig some up during my spring clean in the garden.

Culinary uses: Due to their bitter taste they are best boiled and left in the cooking water overnight before use. They can then be fried or added to salads. The most common method is to preserve them in oil and vinegar (see recipe)

Medicinal uses: It is an appetizer and diuretic. It can also be crushed and used as a poultice on reddened skin.

Recipe: Sweet and Sour Lampascioni. Clean, trim and pierce Lampascioni bulbs and boil in plenty of salted water with a little vinegar until soft. Drain and then crush them a little. Dress them with olive oil, red wine vinegar, salt and pepper. Serve as an appetizer or on top of pureed broad beans.

Link:
http://www.pfaf.org/user/Plant.aspx?LatinName=Muscari+comosum

Jerusalem Artichoke (*helianthus tuberosus*)

Description: Originally a native to North America, this plant is now widely naturalised throughout Europe. It is a relation of the sunflower and looks similar, growing up to 2.5m tall with bright yellow flowers in late autumn. It is not related to artichokes, but the flavour of the  tubers is reminiscent. Jerusalem is said to be a bastardisation of the Italian word girasole, which means sunflower. Also known as sunroot, sunchoke or topinambour.

Where: Rich damp thickets and in ditches along waysides.

When: The tubers are best harvested in autumn through to early winter, by digging up the whole plant and rummaging for the 5-15cm long tubers, which are usually pinkish in colour.

Culinary uses: The tubers are delicious raw, cooked or roasted. It can be used in the same way as potatoes. Later in the winter it can be thinly sliced and added to salads raw. Try roasting with some garlic, rosemary and olive

oil. The tubers can be, but don't have to be peeled. They are rich in inulin, a starch which cannot be digested by the body. This makes it a good food for people trying to loose weight as it is satisfyingly filling without having many calories. Some people will suffer from flatulence as a result though.

Medicinal uses: It is a folk remedy for diabetes and rheumatism. In fact the inulin can be converted to sugars safe for diabetics. It is also said to be an aphrodisiac.

Recipe: Jerusalem artichoke Soup: Chop an onion and 2 sticks of celery in some butter until soft, adding garlic and salt towards the end. Add a kilo of Jerusalem artichokes and a litre of chicken stock. Bring to the boil and simmer for 45-60 mins. Until chokes start to break down. Puree soup until smooth, season to taste and serve with some sage croutons.

Caution: Some people are intolerant to inulin and may experience digestive problems and discomfort after consumption. If unsure try small quantities initially and monitor the effects over the next 12 hours.

Link:
http://www.pfaf.org/user/Plant.aspx?LatinName=Helianthus+tuberosus

Judas Tree (*cercis siliquastrum*)

Description: A small deciduous tree displaying a riot of deep pink flowers in spring, followed by small heart-shaped leaves and finally bean-like seed pods. It is part of the legume family and the flowers and pods are reminiscent in shape of those of broad beans.

Where: Parks and arid woodlands in southern and south-eastern Europe.

When: Flowers in spring and pods appearing in late summer.

Culinary uses: The flowers make a great addition to the salad bowl with their sweetish acid flavour. Buds can be pickled and used like capers. According to one source, the seed pods are also edible raw, but not the seeds.

Medicinal uses: None known

Link:
http://www.pfaf.org/user/Plant.aspx?LatinName=Cercis+siliquastrum

Juniper (*juniperus*)

Description: A genus of coniferous plants of the cypress family. They range widely from low shrubs to large trees with needle-like leaves. The berries are not true berries, but small tight cones. They are about pea-sized and purple in colour when ripe.

Where: The various species each have their preferred habitat all over the northern hemisphere, from j. communis in the Highlands of Scotland to j. oxycedrus on dry rocks in southern and south-eastern Europe. The one pictured was taken along the Via Francigena in Italy.

When: 'Berries' ripen in late autumn.

Culinary uses: Berries are most commonly known as flavouring for gin. In Italy a liqueur is made called gineprino. A couple of juniper berries added whole to game, roasts or stews add an aromatic quality. However they should not be eaten as in concentration the flavour is unpleasant. It is also an ingredient in pickling spice (see recipe below).

Medicinal uses: The fruit of j. communis is the most effective species in herbalism. It is used as a treatment of urinary and digestive disorders as well as arthritis, gout and rheumatism.

Recipe: Pickling spice: Combine the following ingredients: 2 sticks cinnamon, broken up, 1 tsp each of: mustard seeds, black peppercorns, cloves, juniper berries, ground mace, dill seeds, dried ginger, coriander seeds and 4 bay leaves. Crush lightly and boil briefly in vinegar before pickling gherkins or whatever else you like to pickle.

Caution: Large quantities are toxic and can cause renal damage. Should not be used in quantity by pregnant women.

Link:
http://en.wikipedia.org/wiki/Juniper

Lamb's Quarter (*chenopodium album*)

Description: Also known as goosefoot or fat-hen, this annual green grows in an erect fashion up to about 1 metre. The leaves alternate, are toothed and roughly diamond shaped. Flowers appear in summer and are small and green and appear as spikes on the top of the plant. I have chosen this common name over fat-hen (more commonly used in the UK) as the latter is also used for another unrelated species known as smearwort (Aristolochia rotunda), which is not edible.

Where: Abundant on open fertile ground and old manure heaps. Often the first weed to emerge on freshly turned soils.

When: Spring to summer

Culinary uses: Cooked they make an excellent spinach substitute. Combined with beans they are said to lessen wind and bloating. Seeds can be ground and combined with flour in bread. Seeds can also be sprouted and added to salads. It is a healthy and nutritious addition to the diet, although it can be mildly toxic, see caution.

Medicinal uses: An infusion of the leaves is used in the treatment of rheumatism. They can also be applied externally as a poultice to insect bites, rheumatic joints and swollen feet as well as to the head in case of sunstroke. Chewing the seeds is said to relieve urinary problems. The juice of the stems can be applied to sunburn or freckles.

Caution: The leaves and seeds of this plant contain saponins. Whilst toxic, they are poorly absorbed by the body and are also largely broken down during the cooking process. Many varieties of beans also contain saponins. It is therefore recommended that lamb's quarter is best consumed cooked and in moderation.

Link:
http://www.pfaf.org/user/Plant.aspx?LatinName=Chenopodium+album

Leek, Wild (*allium rotundum*)

Description: There are a number of different allium species I collect, some looking like thin leeks, some like spring onions and others like chives, not to mention the garlic varieties. The best of them is probably this species of wild leek.

Where: Disturbed fields, cultivated land, roadsides.

When: Spring

Culinary uses: Raw in salads, in stir fries, added to omelettes and stews and any other way you would normally use leeks. Eat the bulb and the lower green part, peeling off some of the outer leaves.

Medicinal uses: All alliums boost the immune system and are effective to help combat colds and flues. They combat mouth inflammations, infections, hypertension, dyspepsia and arteriosclerosis.

Link:
http://server9.web-mania.com/users/pfafardea/database/plants.php?Allium+scorodoprasum+rotundum

Lemon Balm (*melissa officinalis*)

Description: This perennial herb grows in clumps up to 60cm tall and looks somewhat like the stingy nettle. The delicate leaves exude a distinct lemon aroma when rubbed.

Where: Moist ditches, wastelands, often interspersed with mint or nettles

When: The plant dies back in winter. Leaves can be used any time between spring and autumn.

Culinary uses: A bunch of leaves added in salads add a pleasant lemon flavour. Add to a soup or stew at the end of the cooking process. Stuff a whole bunch into a fish when roasting. A pesto (see recipe) can be made from the leaves or it can be added to a herb butter. Finally it makes a pleasant tasting tea and a sprig can be added to a cold summer drink.

Medicinal uses: It is traditionally known to have a calming effect on the nerves taken as a tea. It raises the spirits when feeling a little depressed. It is effective in the treatment of cold sores. It also relieves fevers and colds. Externally it is used to treat insect bites as well as being an insect repellent.

Recipe: Lemon balm pesto: Combine a generous handful of fresh lemon balm leaves with 4 cloves of garlic and a good helping of good olive oil in a food processor. Serve with spaghetti.
Chicken Breast on a Bed of Wild Herbs with Roast Potato and Fennel: In the centre of an oiled fireproof dish arrange whole bunches of wild herbs such as lemon balm, fennel greens and oregano. Lay chicken breast on top, arrange cubed potatoes and roughly chopped fennel bulb around the edge and sprinkle everything with chopped wild chives and olive oil. Season and bake for 40 minutes until done.

Link:
http://lemonbalm.org/

Lettuce, Wild (*lactuca*)

Description: There are a bewildering number of relations to our common cultivated lettuce growing in the wild, in fact there are some 100 species. Amongst the most common are l. viminea, which looks very much like the sow thistle and Prickly Lettuce (l. serriola). L. virosa is valued for its medicinal properties. Heights and appearance vary considerably and a future project might be a guide to wild lettuces. The thing they all have in common are that they form heads in panicles of yellow, brown or purple flowers in ray florets. Most are edible, although more bitter than the cultivated varieties, some in fact too bitter to be palatable.

Where: Most like open grasslands, walls or rock crevices.

When: Usually best in spring.

Culinary uses: Eat raw in salads or some varieties can be cooked as a potherb.

Medicinal uses: As bitter herbs they all have a positive effect on the digestion and act as a tonic. The latex of l. virosa has a sedative and mildly narcotic effect and is an aid to combat insomnia.

Caution: Some of the more bitter varieties, including l. virosa contain low levels of toxins and should be used with caution and in moderation.

Link: http://en.wikipedia.org/wiki/Lactuca

Locust, Black (*robinia pseudacacia*)

Description: This fast growing, at times invasive, deciduous tree is native to North America, but is now naturalised all over Europe. It grows to a height of 25m. The young stems and branches arm themselves with 2-3cm long, needle-sharp thorns. The foliage is arranged in pairs of simple oval shape. The grape-like white flower-bunches fill the air around them with a sweet fragrance. It is also sometimes known as False Acacia.

Where: Open woodlands

When: The flowers in spring and the seeds and young pods in summer.

Culinary uses: In Italy and France the flowers are eaten as fritters. The seeds and young pods can be boiled like peas. In fact they must be cooked, see notes on caution.

Medicinal uses: Cooked flowers are eaten as a treatment for eye ailments. They also contain an anti-tumour compound. The leaf juice inhibits viruses.

Caution: All parts of the plant except the flower are toxic. The toxin is broken down by heat, so cooked seeds and pods are safe to eat.

Link:
http://www.pfaf.org/user/Plant.aspx?LatinName=Robinia+pseudoacacia

Mallow (*malva*)

Description: A genus of about 25-30 species, which are all edible. They have the characteristic palmate lobed leaves in common as well as the mauve flowers (the name of the colour comes from the French name for the plant) with typically 5 petals. The seeds are sometimes known as cheeses because of their round flat shapes. The roots of the sub-species marshmallow were once used to make the popular confectionary.

Where: Waste-grounds, moist ditches, along pathways and hedgerows. M. sylvestris is common all over Europe.

When: Spring through to autumn

Culinary uses: Young leaves and flowers can be added raw to salads. Leaves cooked make a great addition to soups, where they act as a thickener. The seeds can be nibbled raw as a snack.

Medicinal uses: The whole plant is rich in mucus and a tea made from the leaves, flowers and/or the roots relieves coughs and other chest complaints. The leaves make a good laxative for children and has a calming effect on stomach ulcers. The leaves and flowers can be used as a poultice on bruises, burns, inflammations and insect bites.

Recipe: Malokhia Egyptian Soup: This is my version of this popular recipe. Crush some garlic, whole coriander seeds and dried chillies to taste together and heat gently in some olive oil for about a minute until the aromas are released. Add a couple of handfuls of finely chopped mallow leaves, stir and then add about a litre of chicken stock. Simmer for 10 minutes, season to taste and serve over some rice. Other vegetables of your choice can be added together with the mallow.

Caution: Avoid if grown on cultivated land which may have had inorganic nitrogen fertiliser added

Link:
http://www.pfaf.org/user/Plant.aspx?LatinName=Malva+sylvestris

Marigold, Field (*calendula arvensis*)

Description: A low growing pretty annual flower. It grows to about 15cm in height and has lance-shaped leaves from a slender, hairy stem. The up to 4cm single flower heads are bright yellow-orange.

Where: Native to Southern Europe, but found further north as a garden escapee, it is common on sunny banks, along waysides and in vineyards.

74

When: Spring to summer

Culinary uses: The leaves and young shoots can be eaten raw or cooked. They are rich in vitamins and minerals. The immature flower buds can be pickled like capers.

Medicinal uses: Field Marigold has similar properties to Pot Marigold and has versatile uses in herbal medicine. Externally it is used to heal a variety of skin problems as well as cuts, sprains, stings, wounds, sore eyes and varicose veins. It also has similar properties to Arnica. Internally it is used as a de-toxifying herb and to treat fevers and chronic infections.

Recipe: Marigold Buds under Oil: Bring 1l of white wine vinegar and 1l of white wine to the boil and add some salt. Add 1 kg of cleaned marigold buds (use dandelion buds as an alternative or in combination) in handful batches to the boiling mixture for a minute at a time and fish out with slotted spoon and dry off on some kitchen paper. Slice 4-5 cloves of garlic and distribute in glass jars. Pack in cooked buds and top up with good quality olive oil. Seal and consume as an antipasto or condiment. It will keep like this for up to a year.

Marigold Skin Healing Cream: Heat 150g dried marigold flower heads with 500g lard until melted and well combined. Leave to cool and rest for 3 days. Heat again and filter hot through muslin. Apply to most skin condition and light injuries.

Link:
http://www.pfaf.org/user/Plant.aspx?LatinName=Calendula+arvensis

Marjoram and Oregano (*origanum majorana, o. vulgare*)

Description: These two herbs are so closely related that they can be described together. In appearance they differ in that marjoram has small, grey-greenish leaves, whilst oregano has firmer, slightly larger, light green leaves. Marjoram is a compact, upright, shrubby plant up to 45cm tall, whereas oregano has a dense, spreading habit and grows up to 60cm tall. Most sources will tell you that oregano has a stronger flavour and aroma, however the sub-variety sweet marjoram is sweeter and spicier than both pot marjoram and oregano. Both leaves and dried flower heads are used.

Where: On meadows, wastelands and along pathways.

When: Summer

Culinary uses: Widely used as flavouring in salads, stews, pasta sauces, pizza and generally in Mediterranean cooking. A herbal tea can be made from either variety. They are used in herb mixtures such as Bouquet Garni or Herbes de Provence.

Medicinal uses: A tea made from marjoram aids digestion and relieves flatulence. Both marjoram and oregano teas help relieve the symptoms of colds and coughs, headaches and menstrual pains. Leaves of both are anti-septic and can be applied externally to swellings, rheumatism and stiff necks.

Link:
http://www.herbsociety.org/factsheets/oregano.pdf

Mint (*mentha*)

Description: The mints are a versatile family coming in a great variety of colours, textures, scents and tastes. What they all have in common is the basic leaf shape and all are highly aromatic. They are perennial but mostly die back in the winter

Where: Most prefer moist, semi-shaded locations. They are common throughout Europe.

When: Spring to autumn

Culinary uses: Mint is one of the best known and versatile herbs. Here just some of the uses: a mint sauce for meat dishes, some fresh chopped mint on new potatoes, on salads, Greek style yoghurt and mint sauce, tea, flavouring sweet dishes, mint ice-cream, the list goes on.

Medicinal uses: Mint aids digestion. A cup of mint tea after a meal assists those suffering from indigestion. It has anti-oxidant properties and acts as a cancer preventative. A tea relieves symptoms of cold and flu. Its a mild sedative and calms frayed

nerves. It refreshes bad breath. Macerated stems are used as a poultice on bruises and sprains. It is antiseptic.

Recipe: Tsatsiki: Blend together natural, Greek style yoghurt with finely chopped mint, garlic, grated cucumber and a squirt of lemon juice. Season with salt and pepper and serve as a dip, on potatoes or inside a kebab.

Caution: Pennyroyal, which is a low growing, small leafed variety of mint, has been known to cause abortions and can be damaging to the liver if taken in large quantities.

Link:
http://www.helpwithcooking.com/herb-guide/mint.html

Mugwort (*Artemisia vulgaris*)

Description: A perennial herbaceous weed which spreads aggressively on cultivated ground. From a woody root system it reaches an average height of 1.5m on a reddish stem with pointy serrated leaves which are dark green on the upper side and display a whitish downy underside. When rubbed it gives off a warm herbal aroma reminiscent of camphor or chrysanthemums. Alternative common names include Felon Herb or St. John's plant.

Where: Common all over temperate Europe along waysides, as a weed on cultivated land, hedge banks and waste grounds.

When: Spring to summer before flowering. Roots in late summer to early autumn.

Culinary uses: Young leaves added in small quantities to fatty foods, such as eel, duck, goose or mutton, are said to aid digestion and add a bitter-aromatic note. Older leaves become quite unpleasantly bitter. In Germany it is traditionally added to goose (see recipe). It is sometimes used in Chinese, Japanese and Korean cooking. In Korea it is added to rice dumplings to give them their green colour.

Medicinal uses: Use 1-2 tsp of dried leaves per cup to make a tea taken against indigestion. This can also be used to relieve menstrual problems

and combat intestinal worms. The dried roots harvested in late summer / autumn are even more effective. The compressed dried leaves and roots are also used in the Chinese medicine therapy of moxibustion. A leaf placed inside the shoe is said to help sore feet.

Recipe: Goose with apple and mugwort stuffing: Clean and gut one whole goose and rub with salt and pepper. Mix together 2 chopped apples, 2 chopped onions and a tablespoon of finely chopped mugwort. Stuff the mixture into the cavity and cook in a covered casserole pan for 90 minutes, occasionally basting it with beer and some of the cooking juices.

Caution: Mugwort contains thujone, which is toxic if consumed regularly and in large quantities. Consume with moderation and it should be avoided by pregnant women.

Link:
http://www.botanical.com/botanical/mgmh/m/mugwor61.html

Mullein (*verbascum densiflorum*)

Description: The verbascum genus are also known as velvet plants due to the furry texture of their leaves. They initially form a rosette of leaves at ground level and subsequently sending up tall yellow flowers up to 1.5m tall.

Where: Dry, sunny locations along pathways and on quarries preferring chalky soils.

When: Summer

Culinary uses: Flowers can be used to flavour liqueurs.

Medicinal uses: A tea made from the flowers as well as fennel seeds, mallow and sweet violets to help clear chest infections and raspy coughs. It is used in herbal tobaccos for the same purpose. Leaves and flowers contain skin softening mucilage and can be used to reduce eczema and help

heal wounds. The seed oil soothes chilblains and chapped skin. Woolly leaves can be used as emergency wound dressing. An infusion of the flowers in olive oil is used to treat ear infections. According to one source it also cures the use of evil language and the thinking of evil thoughts. I cannot vouch for the latter uses.

Caution: For internal use strain tea through a coffee filter to remove fine hairs which are an irritant.

Link: http://www.ashtreepublishing.com/Book_City_Herbal_Mullen.htm

Mustard, Black (*brassica nigra*)

Description: An annual member of the cabbage family with rough leaves covered in small hair. The flower emerges on stems up to 1.5m tall and consists of 4 small yellow petals. Seeds appear later in up to 2cm long pods.

Where: Sunny locations on poor soils often near the sea.

When: Spring to summer

Culinary uses: Leaves can be finely chopped and used to add a spicy flavour to salads or cooked as a potherb. Immature flowering stems can be cooked like broccoli. The seeds are popular in Indian cooking. Usually they are added to hot oil before anything else and allowed to 'pop' to release all their aromas and flavours. It's also used in pickling spice (see juniper).

Medicinal uses: Hot water poured on bruised seeds makes a stimulant foot bath, good for colds and headaches. The seed is also used internally as an appetizer, digestive, diuretic, emetic and tonic. A decoction of the seeds is used in the treatment of indurations of the liver and spleen.

Link:
http://www.naturalmedicinalherbs.net/herbs/b/brassica-nigra=black-mustard.php

Mustard, Garlic (*alliaria petiolata*)

Description: A biennial flowering plant in the mustard family growing up to 1 metre tall with heart shaped slightly wrinkled leaves, delicate clusters of white flowers which turn into small upright fruits that release seeds in mid-summer. Also known as Jack-by-the-hedge. When crushed all parts of the plant have a distinct aroma which is a cross between mustard and garlic.

Where: Common along damp hedgerows and woodland edges.

When: Spring for the leaves and flowers, summer for the seeds.

Culinary uses: The young leaves and flowers give salad an agreeable spicy edge. The chopped leaves mixed with some cream cheese make a delicious sandwich filling. Cooked they can be added to spring soups and go particularly well with fresh peas. A white herb sauce with garlic mustard is great on top of new potatoes. Crush the seeds with salt as a condiment.

Medicinal uses: Consuming garlic mustard aids digestion. It is rich in vitamin C. A decoction is used in the treatment of bronchitis and to promote sweating. The roots are also used externally, heated in oil and then applied to the chest against chest complaints. Externally they are used as a poultice on ulcers and to relieve itching and bites.

Recipe: Garlic Mustard Sauce: Make a béchamel sauce by melting some butter over a low heat. Stir in flour and stir continuously for a minute or so. Gradually add milk, always stirring, until a rich creamy consistency is achieved. Season with salt, pepper and nutmeg and add a good handful of chopped garlic mustard leaves. Leave to infuse for a minute and serve over new potatoes or on fish.

Link:
http://www.pfaf.org/user/Plant.aspx?LatinName=Alliaria+petiolata

Myrtle, Common (*myrtus communis*)

Description: An evergreen shrub growing up to 5m tall with simple shiny, almost leathery leaves displaying small white flowers in the spring and dark purple-blue berries in late autumn the size of a large pea containing several seeds. All parts of the plant are highly aromatic.

Where: Sunny hedges and woodland edges around the Mediterranean.

When: Leaves all year, flowers in spring and berries in late autumn.

Culinary uses: Most famously used as ingredient for mirto liqueur from Sardinia or Corsica. Mirto Rosso is made from the berries, whilst the leaves make Mirto Bianco. Flowers can be added to salads. Leaves used to flavour meat dishes and casseroles in the same way as bay leaf. The berries can be eaten raw or cooked into a jam (see recipe under autumn olive). The dried berries are also used as a flavouring in a similar way to juniper berries. A tasty tea can be made from dried leaves and berries.

Medicinal uses: The tea or a steam bath is excellent to clear the airways and a safe way to treat sinusitis, bronchial congestion and dry coughs. The essential oil is used in treating cold sores. It has a great anti-bacterial action making it effective in the treatment of urinary infections, digestive problems and vaginal discharge. Externally it is used to treat acne, gum infections and haemorrhoids.

Link:
http://www.pfaf.org/user/Plant.aspx?LatinName=Myrtus+communis

Navelwort (*umbilicus rupestris*)

Description: Also known as wall-pennywort, named for its shape resembling a human navel. A perennial succulent plant up to about 5cm in diameter, rarely up to 10cm. The flowers appear in summer on little tightly clustered fleshy stems.

Where: On walls and rock crevices all over Europe to as far north as Scotland and as far south as North Africa.

When: Best in early spring

Culinary uses: They are delicious and refreshing popped straight into the mouth off the wall with their crunchy texture. They make a good addition to the salad bowl. By summer when they start turning a tinge of red they become unpleasantly bitter.

Medicinal uses: They have a cooling effect whichever way they are used. Externally as a poultice they offer relief from burns, inflamed areas of the skin, acne and piles. Internally the juice is diuretic and helps with infections of the liver and spleen. The juice is also said to relieve earaches. The plant has a folk reputation to help combat epilepsy, but I have not noticed any improvements having used it on an epilepsy patient.

Link:
http://pathtoselfsufficiency.blogspot.com/2012/02/navel-wort.html

Nettle, Red Dead (*lamium purpureum*)

Description: An annual, low growing nettle variety, growing only up to 25cm, with soft downy, non-stinging leaves with a purplish-red tinge on the tip of the plant and delicate light purple flowers.

Where: Moist ditches and meadows everywhere.

When: summer

Culinary uses: Young leaves and shoots raw in salads or cooked as a potherb. The flavour is somewhat bland and is therefore best mixed with other greens.

Medicinal uses: Bruised fresh leaves can be applied externally to cuts and bruises

Link:
http://www.arthurleej.com/a-deadnettle.html

Nettle, Stinging (*urtica dioica*)

Description: One of the most recognised and common weeds. It is a coarse perennial growing in clumps from rhizomes. The serrated leaves have stinging hairs and it displays tiny green flowers in summer. Needless to say, leaves are best harvested using gloves.

Where: Everywhere alongside paths, waste-grounds and ditches, preferring a slightly alkaline soil.

When: Almost all year.

Culinary uses: Young leaves only can be cooked as a potherb and make an excellent addition to a soup. Nettle leaves are very nutritious and contain high concentrations of iron as well as other vitamins and minerals. A warming tea can be brewed from the dried leaves. To add flavour add to China tea. The stinging effect is neutralised by heat or thorough drying.

Medicinal uses: Nettles are valuable as a blood purifier and cleansing tonic. An infusion helps treat internal bleeding, anaemia, excessive menstrual bleeding, arthritis and rheumatism. It stimulates circulation and clears uric acid. It also reverses prostate enlargement. Externally it is used in the treatment of eczema and haemorrhoids.

Caution: Avoid old leaves as they can be an irritant to the kidneys if eaten.

Link:
http://www.ageless.co.za/herb-stinging-nettle.htm

Nettle, White Dead (*lamium album*)

Description: This perennial non-stinging nettle's long rhizomes bear erect stems, up to 60cm in height, that are square and hollow, with opposite pairs of bright green leaves and clusters of tubular white flowers.

Where: Waste grounds, damp ditches, meadows, roadsides.

Where: Waste grounds, damp ditches, meadows, roadsides.

When: Most of the year, but leaves best used in spring.

Culinary uses: Young leaves raw in salads. Leaves cooked as a potherb. A tea can be obtained from the flowers.

Medicinal uses: It narrows blood vessels and is thus used in the treatment of excessive menstrual bleeding. It is generally regarded as a tonic for the reproductive organs. Dried leaves sniffed can stop nose bleeds. An infusion is effective in the treatment of kidney, bladder and prostate complaints as well as diarrhoea. Externally it can be applied to piles and haemorrhoids.

Link:
http://www.naturalmedicinalherbs.net/herbs/l/lamium-album=white-dead-nettle.php

Oak (*quercus*)

Description: There are some 600 species of oak trees. They can be deciduous or evergreen and are common all over the northern hemisphere. Most species can be recognised by the classic lobed leaf shape and the acorn fruit born in autumn. Q. alba the white oak, is probably the most common oak in Europe

Where: Woodlands

When: Autumn for the seeds

Culinary uses: The seeds (acorns) of most species are to a greater or lesser extend decidedly high in tannins and therefore bitter. For consumption they should first be dried, then ground and 'leached' This is achieved by thoroughly washing and soaking them in water. Boiling and changing the water several times also has that effect, however many of the useful minerals are also lost. Treated like this it can be used as a flour substitute. Try making bread with part acorn flour. Simply roasting and grinding the seeds makes for a good coffee substitute. Young leaves make an interesting country wine.

Medicinal uses: The seed and bark is used for its antiseptic and astringent qualities. The bark boiled and the liquid drunk is used in the treatment of diarrhoea, fevers, coughs and colds, asthma and lost voice. The bark chewed helps mouth sores. Externally the bark can be used as a wash for burns, rashes, bruises, ulcers and as a vaginal douche.

Link:
http://www.pfaf.org/user/Plant.aspx?LatinName=Quercus+alba

Old Man's Beard (*clematis vitalba*)

Description: Also known as Traveller's Joy, it is a perennial, deciduous climbing shrub growing up to 15 metres with branched grooved stems, forming greeny-white, scented, flowers, which go fluffy giving it its common name of Old Man's Beard.

Where: Woodland edges and hedgerows.

When: Early spring for the young shoots

Culinary uses: Only the outer tips of young shoots are edible and can be eaten cooked like hop shoots or asparagus.

Medicinal uses: The juice of the plant applied to the inside of the nostrils is used to relieve migraine attacks, but can also destroy mucous membranes.

Caution: All parts of this plant are poisonous, however the toxicity is dissipated by cooking or drying. Only use the very tips of the young shoots and don't over-indulge.

Link:
http://www.pfaf.org/user/Plant.aspx?LatinName=Clematis+vitalba

Orach, Hastate or Halberd-Leaved (*atriplex hastata*)

Description: An annual erect plant growing up to 80cm tall with almost triangular leaves.

Where: Near the coast, on dunes above the tide mark, mostly northern and Eastern Europe. It is protected in some places.

When: Spring for the leaves

Culinary uses: The leaves make an acceptable spinach substitute. The dried seeds ground to a flour can be added to bread, but they are very fiddly.

Medicinal uses: None known.

Link:
http://www.pfaf.org/user/Plant.aspx?LatinName=Atriplex+hastata

Ox-tongue, Bristly (*picris echioides*)

Description: An annual or biennial weed with a rosette of lanceolote leaves up to 20cm long at ground level from which a flower stem of up to 90cm emerges with yellow dandelion-like flower heads. Both the leaves and stem are distinctly bristly.

Where: Waste grounds and meadows in Southern Europe, but also naturalised further north.

When: spring and summer.

Culinary uses: Young leaves cooked and mixed with other greens as they are somewhat bitter.

Medicinal uses: None known

Link:
http://www.pfaf.org/user/Plant.aspx?LatinName=Picris+echioides

Parsley, Wild or Cow (*anthrisiscus sylvestris*)

Description: A common roadside perennial herb growing up to 1.2m tall. The leaves look like common flat-leafed parsley, but they do not emit much aroma unlike the cultivated variety. However, the long, tapering roots smell strongly of parsley. The small white umbellate flowers appear in summer.

Where: Roadsides and open fields and meadows, also in shady locations.

When: Spring to autumn.

Culinary uses: The leaves can be used raw, but develop more flavour when cooked in a stew or casserole. The roots have more flavour and can be cooked as a vegetable or to impart flavour to a stew.

Medicinal uses: Soak the roots in rice water for several days and then cook with other vegetables as a tonic for general weakness.

Caution: Do not confuse with fool's parsley aethusa cynapium or poison hemlock conium maculatum, which are both poisonous and can cause pain, confusion of vision and vomiting. The leaves of fool's parsley are finer and more delicate to those of true parsley, whilst hemlock's lower stem is usually spotted or streaked red or purple and when crushed the leaves and roots emit a rank, unpleasant smell.

Link:
http://www.pfaf.org/user/Plant.aspx?LatinName=Anthriscus+sylvestris

Pellitory of the Wall (*parietaria officinalis*)

Description: This is one of those plants I walked past hundreds of times before I realised how useful it was. It is a perennial member of the nettle family it has smaller leaves than the stingy nettle growing off red stalks creeping along walls. The flowers are similarly inconspicuous to those of the stinging nettle.

Where: Mainly in cracks along old stone walls all over Europe except Britain and Iberia.

When: Spring to summer.

Culinary uses: The leaves and young shoots make an excellent vegetable added to soups (see recipe), mashed potato or a risotto.

Medicinal uses: This plant has long been valued as an excellent diuretic, increasing urine production to clear the system. Combined with stingy nettle the effect is increased. Drink an infusion of 2 tablespoons of dried herb per litre of water 3 times a day or make an infusion and take 40 drops three times a day. This is helpful to combat metabolic illnesses, including obesity, diabetis and cellulites, rheumatic illnesses, such as gout and arthritis, illnesses of the urinary tract, including gallstones and kidney stones as well as renal inflammations and cystitis and illnesses of the circulatory system, such as high blood pressure and water retention. Externally is used as a disinfectant wound dressing.

Recipe: Pellitory of the Wall Soup: Boil a large bunch of leaves for some 10 minutes in salted water. Drain and puree. In a separate pan heat up a little butter, add 50g of flour and 400ml of milk and bring to the boil slowly. Add the pureed Pellitory and season to taste.

Caution: The pollen are responsible for symptoms for many hay fever sufferers. If you have a known allergy, avoid contact with the flowering plant.

Link:
http://www.botanical-online.com/medicinalsparietariaangles.htm

Periwinkle, Greater (*vinca major*)

Description: This perennial shrub-like creeping vine grows up to 50cm tall and individual plants can cover an area of up to 2m in diameter as groundcover. The 5cm long leaves are pointy with a shiny surface and the flowers with their 5 petals are amongst the truest blue shade along with borage flowers. This plant is only of passing interest for the forager for its medicinal properties.

Where: Hedgerows and the edges of cultivated land. Native to southern Europe but naturalised in other areas too including Britain.

When: Spring to autumn.

Culinary uses: None

Medicinal uses: Internally it is used to reduce menstrual bleeding. It is given to treat ulcers and sore throats and to reduce blood pressure. Externally it is also used as a wound herb and to stem a nose bleed. It should be administered by a qualified herbalist.

Caution: Large doses of this plant are poisonous.

Link:
http://www.pfaf.org/user/Plant.aspx?LatinName=Vinca+major

Pigweed (*amaranthus albus*)

Description: Pigweed is one of approximately 60 species of the amaranth genus, all of them are edible. It is one of my favourite plants as it is the anarchist of the plant world. It is resistant to most weed killers throwing a spanner in the works of the GM crop lobby. Genetically modified food crops are bred to resist their own brand weed killer, but they have not found one that would kill pigweed whilst sparing the food crops. In addition the Incas and Aztecs used it as major food crop for both its seeds and leaves. The conquering  Spaniards banned it though because the seeds were baked to cakes with some human blood for religious rituals, which the Spaniards did not approve of. The various members of the amaranth vary considerably in appearance. Pigweed, the most common, is an annual growing up to 1.5m tall with matt, light green leaves, usually with a whitish pattern on it. The flower is upright, candle-like or feathery, containing masses of small black seeds. Other species can vary considerably in colour and seed size.

Where: Disturbed ground anywhere

When: Summer to autumn.

Culinary uses: Young leaves can be used in salads, all leaves, as long as they are healthy looking, can be cooked like spinach. The leaves are rich in minerals and vitamins, but do not have much of a flavour. The seeds can be ground and used as flour. Cooked whole they become gelatinous, but need to be chewed very well, else they will pass straight through the system without giving off its goodness. They are very high in protein and some essential amino acids. However for most varieties the seeds are quite fiddly to collect in quantity.

Medicinal uses: None known

Link:
http://en.wikipedia.org/wiki/

Pine, Stone (*pinus pinea*)

Description: The seeds of all pine trees are in theory edible, but only the seeds of the Italian stone oak are of sufficient size to make them worth gathering. The tree grows to a considerable height with long, evergreen needles. The cones are the size of a large fist and the up to 1cm long seeds are hidden behind the scales and encased in a hard nutshell. It's a fiddly and messy job to first release the seed shells, then crack the tough outer casing to get to the tasty seed. It explains why they are so expensive when bought, but  the flavour of freshly gathered seeds far exceeds that of commercially bought ones.

Where: Mediterranean woodlands

When: Early autumn

Culinary uses: The seeds, once you get to them, make a delicious snack. Together with olive oil, basil, garlic and Parmesan cheese they are a main ingredient in traditional Pesto Genovese. It also adds a nice crunchy texture to many pasta sauces and is good in sweet dishes such as ice-cream or cakes.

Medicinal uses: The turpentine obtained from the resin of all pine trees is used in the treatment of bladder and kidney infections. Internally and externally as a rub or steam-bath it relieves rheumatism. It is also beneficial for a number of respiratory and skin complaints.

Recipe: Pesto Genovese: Combine in a food processor or pestle and mortar: a generous handful of fresh basil, a handful of grated Parmesan cheese, a small handful of pine kernels, 1or 2 gloves of garlic and a good dollop of olive oil and work to a paste. Serve on pasta.

Link:
http://www.pfaf.org/user/Plant.aspx?LatinName=Pinus+pinea

Plantain, Ribwort or Common (*plantago lanceolata or p. major*)

Description: The two most common species of the plantago family (nothing to do with the banana) are Ribwort (photo right) and common plantain (photo below). Both are more useful as medicinal plants and have similar properties. However there is a third less common species, Buck's Horn Plantain (p. coronopus, photo bottom), which is the best of them for eating. Ribwort Plantain is marked by its narrow pointy leaves, whilst Common Plantain has wider leaves. Both are ground hugging plants and are decidedly  stringy. When the stalk is broken a few threads will show. The flowers appear on thin stalks, starting off green then turning brown as the seeds mature.

Where: Meadows, lawns and waste grounds. Buck's Horn plantain grows on sandy soils near the sea.

When: Almost all year

Culinary uses: Buck's Horn Plantain is best blanched and then added to a salad. The other two are only really used in an SAS type survival situation and young leaves only with the stalks removed and cooked as a potherb. The seeds cooked can be eaten like sago or ground can be added to flour for making bread.

Medicinal uses: This plant is much more valuable as a medicinal herb. A tea made from the leaves, seeds combined with the leaves and flowers and/or roots of dandelion make an excellent liver tonic. A tea made from the leaves and flowers sweetened with honey relieves bronchial complaints. An infusion of the whole plant can be gargled to combat throat and gum infections. The juice applied to insect bites relieves itching. It is also an effective treatment to stem bleeding. Internally it is used for a wide

variety of complaints including diarrhoea, gastritis, peptic ulcers, irritable bowel syndrome, haemorrhage, haemorrhoids, cystitis, sinusitis, asthma and hayfever. The seeds combat parasitic worms.

Link:
http://www.pfaf.org/user/Plant.aspx?LatinName=Plantago+major   &
http://www.pfaf.org/user/Plant.aspx?LatinName=Plantago+lanceolata

Plum (*prunus domestica*)

Description: There are many varieties of wild and cultivated plums which come in a multitude of sizes, and colours, but all are deciduous trees of medium size, bearing yellow, purple, red or green fruit with one single stone in the centre.

Where: Abandoned orchards, hedges and woodlands

When: Summer to autumn, depending on variety

Culinary uses: Fruit is perfect for eating fresh, cooked into jams or stewed, dried as prunes etc. Flowers can be sprinkled onto a salad.

Medicinal uses: Dried fruit is an effective and well-known laxative.

Caution: Seeds and leaves contain hydrogen cyanide which is toxic. Seeds do contain edible oil, but do not use if very bitter tasting.

Link: http://www.pfaf.org/user/Plant.aspx?LatinName=Prunus+domestica

Polypody (*polypodium vulgare*)

Description: A member of the fern family, the fronds with triangular leaflets are arranged in 10-18 pairs and grow up to 50cm tall. The leaflets become much shorter towards the end of the frond.

Where: On walls, rocks and trees in shady moist locations all over Europe

When: Best harvested in late autumn

Culinary uses: The rhizomes contain osladin, a compound 500 times sweeter than sugar. It therefore has found use in confectionary such as nougat or liquorice. On its own it is cloyingly sweet and quickly becomes quite sickening.

Medicinal uses: Both an infusion of the roots or leaves, although the latter is less effective, is used to stimulate bile secretion and has a gentle laxative effect. It is also effective in treating liver diseases. A syrup of the whole plant is effective in dispelling parasitic worms.

Caution: Although no toxicity is reported for this plant, many fern species contain carcinogens and/ or toxins. Some rob the body of the vitamin B complex, but people who are on a varied diet and consume moderately should be fine.

Link:
http://www.naturalmedicinalherbs.net/herbs/p/polypodium-vulgare=polypody.php

Poppy, Corn (*papaver rhoeas*)

Description: There are numerous species within the poppy family, including of course the opium poppy. The most common however is the corn poppy, the edible part being the seeds, which have no hallucinogenic effect. Other parts of the plant are mildly poisonous, although the flower is occasionally used in herbalism. Everyone recognises this delicate two-petaled flower, which can be seen from a distance with its bright red flowers, off-setting them from its surroundings. The seeds are harvested from the up to 2cm long, slightly hairy seed pods when they are thoroughly dry and a slight rattling can be heard inside when shaken.

Where: Fields and meadows everywhere.

When: summer

Culinary uses: The seeds can be added to breads or cereals and on salads. In Germany bread rolls are often covered with poppy seeds. The leaves and flowers can be cooked and eaten as a potherb. The flowers make red food dye.

Medicinal uses: A tea made from the flowers sweetened with some honey helps against colds and to calm hyper-active children. Tea made from poppy is a mild sedative, but unlike the sap of the capsules of the opium poppy, is non-addictive. It is also useful to lower fevers and has anti-cancer properties.

Caution: Also the latex of the immature corn poppy has some narcotic effects, although less so than that of the opium poppy. It can be used as a sleep-inducing sedative, but only under expert supervision. Avoid handling poppy if travelling to the Middle East, as incidences of people being arrested are reported for having minute quantities of poppy seeds attached to their clothing.

Link:
http://www.pfaf.org/user/Plant.aspx?LatinName=Papaver+rhoeas

Primrose (*primula vulgaris*)

Description: A low growing perennial flower with a basal rosette of coarse leaves. The flowers, although in cultivated varieties can be almost any colour, in the wild are generally pale yellow born on slender stalks no taller than 30cm. It's amongst the first flowering plants in spring.

Where: Woodlands and hedgerows or north-facing slopes.

When: Early spring

Culinary uses: Both leaves and flowers are edible. The leaves can be added raw to salads, but tend to be a little coarse in texture. They are better cooked as a potherb. The flowers make a decorative addition to the salad bowl. The flowers can also be used to make an agreeable country wine. My favourite use must be my personal adaptation of a 17th Century English recipe, a spring tart (see below for recipe)

Medicinal uses: In herbalism primroses have long been associated with conditions involving spasms, cramps and rheumatic pains, although Cowslip, p. veris is considered to be more effective in treating these ailments. The plant contains saponins and salysilates, the latter being the main component of aspirin and anodyne and as such has anti-inflammatory properties. An infusion of the roots is a good cure for nervous headaches.

Recipe: Spring Tart: Make a pastry by combining 200g flour, a pinch of salt, 75g sugar, 1 egg and 125g cold butter until you have a smooth, non-sticky pastry. Wrap in foil and leave in the fridge for an hour or so. In the meantime briefly blanch about 1 litre of mixed primrose, sweet violet and wild strawberry leaves, drain and pat dry. Whiz cooked leaves with 400ml cream in food processor and return mixture to the pan and simmer for another few minutes. Take off the heat and stir in 100g grated sponge biscuits, 2 beaten eggs, pinch of salt, ½ grated nutmeg, ½ tsp cinnamon and 30g sugar. Finally add 150ml spinach juice to dye your tart green and a generous handful of primrose and sweet violet flowers, leaving a few aside for the garnish. Roll out pastry to fit into a cake tin with 2cm edges around

the side. Pour in the filling and bake at 175C for 40 minutes. Leave to cool and garnish with the remaining flowers.

Caution: This should not be taken by pregnant women and those with a known allergy to aspirin or those taking coagulant drugs.

Link:
http://www.pfaf.org/user/Plant.aspx?LatinName=Primula+vulgaris

Purslane (*portulaca oleracea*)

Description: A ground-covering succulent plant with up to 1.5cm long leaves, fleshy creeping stalks and finally small yellow flowers.

Where: Open sunny ground and rock crevices mostly in southern Europe

When: Summer

Culinary uses: The succulent leaves and stems are an excellent and tasty addition to any salad. They can also be cooked and make a good thickener in soups, but I prefer the crunchiness it adds raw to a salad with slight salty, minerally and mildly sour character. I also add it to the juicer along with other vegetables such as carrot, tomato and cucumber for a nutritious drink. The plant is extremely nutritious containing, apart from various vitamins (A, B1, B2 and Niacin), the highly sought after essential Omega-3 acids. They are the ones you are normally told are only available in seafood. It is also rich in proteins.

Medicinal uses: Due to its content of omega-3 acids it is useful in preventing heart disease and strengthening the immune system. The fresh juice is used in the treatment of strangury, coughs and sores. A poultice from the leaves can be applied to burns, insect bites and skin diseases. The leaf juice can also be used to relieve earaches.

Recipe: Tomato, Cucumber and Purslane Salad: TexMex recipe best served with grilled seafood. Combine 1 cucumber peeled and chooped, 1 medium chopped tomato, 1 bunch of purslane, chopped, 1 minced Jalapeño chilli,

2-3 tbsp lemon juice and salt and pepper to taste.

Link:
http://www.pfaf.org/user/Plant.aspx?LatinName=Portulaca+oleracea

Raspberry (*rubus idaeus*)

Description: Up to 2m tall perennial, deciduous shrub with woody stems closely resembling its cousin the blackberry, although the branches are less prickly and the fruit is bright pink when ripe.

Where: Woodlands usually in mountainous areas

When: Summer

Culinary uses: Raw or cooked, delicious straight off the bush or in jams, pies and desserts. A herb tea can be made from the dried leaves as with blackberry. It is high in anti-oxidants and rich in vitamins.

Medicinal uses: A tea made from the leaves relieves diarrhoea, and helps with female problems, strengthening the uterus taken in the 3 months before childbirth and relieving painful menstrual cramps. An infusion of the leaves and roots can be gargled as a treatment against tonsillitis and mouth ulcers and as a poultice can be applied to minor wounds, varicose ulcers and burns. Fresh raspberry juice sweetened with a little honey makes a coolant for fevers.

Link:
http://www.raspberrylovers.com/

Rocket, Yellow (*barbarea vulgaris*)

Description: This member of the brassica family is also known as bittercress or herb Barbara. From a basal rosette of shiny dark green, roundly toothed leaves rises a ribbed stem up to 60cm on which dense clusters of yellow flowers are borne.

Where: Moist places along woodland edges, hedgerows and muddy river edges in dappled shade all over Europe.

When: Spring

Culinary uses: Leaves are used raw or cooked and have a hot cress-like flavour. Young leaves chopped make good early season salad additions. They also make a great pesto (see recipe below). Older leaves are cooked as a potherb. Young flowering stems can be cooked like broccoli.

Medicinal uses: As a poultice the leaves have traditionally been applied to wounds (Santa Barbara is the patron saint of miners and artillery soldiers, who routinely used the herb to treat their wounds). The leaves are also rich in vitamin C.

Recipe: Barbara Pesto: Crush a handful of yellow rocket leaves with a little garlic, some toasted pine kernels, some grated parmesan cheese and some good quality olive oil to make a paste. Serve on hot pasta of your choice.

Caution: According to some reports excessive use of this plant may lead to kidney malfunction.

Link:
http://www.pfaf.org/user/Plant.aspx?LatinName=Barbarea+vulgaris

Rocket, Wild or Perennial Wall (*diplotaxis tenuifolia*)

Description: A low growing herb in the mustard family, the leaves are more delicate and more deeply serrated than cultivated rocket and the aroma and flavour are more intensely spicy. Like other members of this family, it sprouts small, bright yellow flowers which form small seed pods later.

Where: Walls and poor, sunny locations mostly in Southern Europe.

When: Spring and summer

Culinary uses: The fresh leaves make an excellent salad addition. Added at the last minute to a pasta sauce it adds spice (see recipe).

Medicinal uses: Rocket is rich in vitamin C and mineral salts and has an appetizing effect.

Recipe: Summer Pasta: Halve some ripe cherry tomatoes and sauté gently in plenty of good olive oil until they are just starting to break up. Stir in a handful of fresh wild rocket, season with salt and pepper and serve on spaghetti topped with some mozzarella cheese.

Link:
http://www.pfaf.org/user/Plant.aspx?LatinName=Diplotaxis+tenuifolia

Rose, Dog (*rosa canina*)

Description: An up to 5m tall deciduous shrub. The branches are covered in vicious thorns, the flowers are made up of 5 white to delicately pink flowers later to be replaced by 2-3cm long hips, which are a rich source of vitamin C.

Where: Hedgerows, woodland edges.

When: summer for the flowers, autumn for the hips, which are best after the first frost.

Culinary uses: Rose petals can be made into a delicately perfumed syrup, jelly or country wine. The hips can be used to make marmalade, jellies, or syrup. They also make an interesting soup. Dried, the hips are used to make tasty and health promoting tea. Rosehips also make a pleasant country wine.

Medicinal uses: Its high content of vitamin C makes rosehip tea an excellent aid against colds and flu and other mild infectious diseases. The syrup can be added to cough mixtures. Apart from vitamin C it is also rich in vitamins A and E as well as flavanoids and essential fatty acids. According to research it may have anti-cancer properties.

Recipe: Hip Soup: Thoroughly de-seed, top and clean a few handfuls of rosehips. Throw into a saucepan of boiling water and boil until tender. Drain into a bowl and force cooked hips through sieve. Measure out 1 1/2 litres of this liquid, adding extra water if needed, return to the saucepan. Add a little sugar and some potato flour or other thickener and bring to the boil stirring constantly. Serve hot or chilled with a garnish of flaked almonds and some whipped cream and a crust of bread.

Caution: Always make sure you remove all trace of the seeds with any recipe as they are an irritant to the bowels.

Link:
http://www.pfaf.org/user/Plant.aspx?LatinName=Rosa+canina

Rosemary (*rosmarinus officinalis*)

Description: An up to 1.5m tall evergreen perennial shrub with almost needle-like highly aromatic leaves up to 2cm in length and a profusion of blue/purple flowers at the base of the leaves for much of the year.

Where: Hedgerows (often in fact grown as a hedge along private properties, but most owners won't begrudge you a sprig or two), dry scrubland.

When: All year

Culinary uses: The leaves finely chopped are part of many Mediterranean dishes, especially casseroles, soups, meat roasts and stews. The whole sprig can be added and removed after cooking as the leaves are somewhat tough. The flower can also be used and has a milder flavour. A fragrant tea is made from the fresh or dried leaves and flowers.

Medicinal uses: Rosemary is said to strengthen the memory and a tea is used to treat headaches and depression. It is also used as a tonic for the kidneys. It is also high in anti-oxidants and a great preventative for heart diseases. Research has also shown that it may be effective in the treatment of toxic shock syndrome.

Link:
http://www.botanical.com/botanical/mgmh/r/rosema17.html

Saffron Crocus (*crocus sativus*)

Description: This autumn flowering purple coloured crocus yielding the most expensive of spices is said to not grow in the wild, however I know of at least one field in the foothills of the Apennines, where they grow freely. They may have been planted deliberately at some stage and then sort of forgotten about. The spice comes from the crimson stigmas. 150,000 flowers and 400 hours of work are needed to produce 1kg of saffron, which accounts for the high price, however luckily only a small quantity is needed to impart flavour and colouring.

Where: Normally just in cultivated fields, but you may be lucky to find an abandoned field.

When: Autumn

Culinary uses: The fine stigmas are used to impart a gentle flavour and bright yellow colour to many dishes such as Paella, Risotto Milanese, Bouillabaisse as well as sweet dishes.

Medicinal uses: Saffron has a long history as a medicinal herb, although is nowadays not used much as such due to its high price and cheaper alternatives being available. The styles and stigmas are useful to induce menstruation, treat period pains and haemorrhages of the uterus. They act as a stimulus and sedative for children and calm indigestion and colic. Commercial saffron is usually adulterated with less expensive ingredients.

Caution: Large quantities of saffron are very toxic, however unless you are very rich you are unlikely to ingest a sufficient amount.

Link:
http://www.pfaf.org/user/Plant.aspx?LatinName=Crocus+sativus

Sage, Wild or Wild Clary (*salvia verbenaca*)

Description: Initially a rosette of oval, roundly toothed leaves with a rough texture. Up to 35cm tall stalks rise from that with beautiful light blue clusters of flowers. The leaves are not nearly as aromatic as the cultivated variety.

Where: On sunny fields and meadows around the Mediterranean.

When: Spring to autumn.

Culinary uses: The leaves only really develop much flavour when cooked and can thus be added to casseroles and soups. The flowers can be added to salads. The young leaves can be eaten fried in butter as a side dish or on top of ravioli. A tea is made from the leaves.

Medicinal uses: The seeds soaked a few minutes in water form a thick mucilage which can be applied to the eye to remove dust particles and other foreign bodies, as with Clary. The tea is said to aid digestion.

Link:

http://www.shootgardening.co.uk/plant/salvia-verbenaca

Saint John's Wort (*hypericum perforatum*)

Description: A perennial plant growing up 1m tall. Pairs of small balsamic scented leaves grow up on an erect, delicate, woody stem. In summer around St. John's Day (24 June) lemon-scented bright yellow flowers appear. If held against the light the leaves display small translucent spots, which are oil glands.

Where: Sunny woodland edges, along hedges and along pathways all over Europe.

When: Summer

Culinary uses: The leaves are used sparingly in salads and to flavour liqueurs. The herb and the fruit can be used as a tea substitute

Medicinal uses: This has a whole plethora of uses. Internally a tea made from the leaves and flowers has successfully been tested in the treatment of mild to medium depression. In clinical tests 2/3 of patients have shown improvement. An extract of the flowers is anti-viral and sedative. Externally it is applied as a poultice to many conditions including cuts, bruises, burns, haemorrhoids and varicose veins. An oil (see recipe below) is applied to the skin as anti-wrinkle treatment, to relieve pains caused by rheumatism or gout, treat burns, improve dry flaky skin, genital ulcers, and is a good insect repellent. It's also been investigated as a possible anti-AIDS drug.

Caution: May reduce the effect of the contraceptive pill and some other medicines. It is also known to cause dermatitis in some people.

Recipe: St. John's Oil: Collect a generous amount of St. John's Wort flowers around mid-June. Stuff tightly into a glass jar and top with soya or olive oil and leave in a sunny spot for some 20 days. Drain and use externally for a number of skin conditions.

Links:
http://www.hypericum.com/hyp09.htm

Salad Burnet (*sanguisorba minor*)

Description: Low growing plant with a rosette of fine stalks bearing 10-12 pairs of delicate, round, serrated leaves each. It grows up to 30cm tall and leaves are up to 1cm across. The flowers are tiny, reddish-pink and almost berry-like born on long stalks.

Where: Very common on grasslands all over Europe

When: Almost all year

Culinary uses: The young leaves have a mild cucumber-like taste and make an excellent addition to the salad bowl. A few leaves added to a summer punch or other fresh drinks are very agreeable. Try mixing with cream cheese as a sandwich filling.

Medicinal uses: Salad burnet is said to lift the spirit and drive away melancholy. The juice or an infusion of the powdered dried roots and leaves can be used both internally and externally to stem bleeding and heal wounds. Taken as a tea it promotes perspiration and has a cleansing effect. It is also said to benefit rheumatism sufferers. Poultices of the leaves are a coolant on sunburns and eczema.

Link:
http://subsistencepatternfoodgarden.blogspot.com/2009/04/dimes-worth-of-salad-burnet.html

Salsify (*tragopogon*)

Description: The tragopodon genus of biennial or perennial plants consists of some 140 species and the common names, Salsify or Goat's Beard, often don't distinguish between them. All have in common an upright habit, branching little and if then upwards. The leaves are grass or onion like. The flowers can be yellow, purple or bronze coloured, the calyx leaves are longer than the flower petals, giving them a star-like appearance.

The seeds are born on large creamy dandelion clocks. The root is a strong taproot. Of greatest interest to the forager are t. porrifolius, growing up to 60cm with leek-like foliage and purple flowers, and t. pratensis, which displays yellow flowers.

Where: Native to the Mediterranean, but naturalised all over Western Europe, it is often found near the coast, on dunes, waste-grounds and meadows.

When: Spring for leaves and young shoots, summer for flowers and late summer / autumn for the roots.

Culinary uses: The roots are most commonly used. Young roots can be grated and eaten raw in salads. Older roots are best cooked and eaten as a vegetable. Their taste is said to resemble that of oysters. The young leaves and shoots are eaten raw in mixed salads or cooked in soups. The flowering shoots are cooked and eaten like asparagus.

Medicinal uses: Both varieties described have a beneficial effect on the liver and gallbladder and may also help to reduce blood pressure. The root of l. pratensis has a high inulin content, which makes it a useful herb for diabetes sufferers.

Recipe: Salsify alla Bolognese: Prepare a tomato and mince meat sauce to your liking. Clean and slice a few salsify roots and gently fry in some olive oil with some salt and thyme. Add the Bolognese sauce and simmer until the root is soft. Serve with pasta of your choice and a grating of Parmesan cheese.

Link:
http://theseedsite.co.uk/profile1011.html

Sea Buckthorn (*hippophae rhamnoides*)

Description: An up to 6m tall deciduous shrub with dense, stiff and very thorny branches. The narrow, pointy leaves display a silvery colour. Female plants produce juicy bright orange berries, about ½ cm in diameter.

Where: Commonly found along seashore, but also planted along canals or as hedging.

When: Autumn for the berries.

Culinary uses: Sea-buckthorn berries are extremely nutritious and rich in vitamins (especially C&E), flavonoids, essential fatty acids and other beneficial nutrients. Eaten raw they are too sour for most palates. Juice as part of a multi-fruit juice. Cook to make jam, pies or steep in alcohol to make a liqueur. In Belgium a beer is made flavoured with sea buckthorn berries. A tea can also be made from the leaves.

Medicinal uses: The health benefits of sea buckthorn would fill a separate book. To list but a few: In infusion of the leaves helps the digestion and intestines and kills intestinal parasites. The fruit has a positive influence on cardiac disorders and has strong anti-cancer properties. Used externally it heals burns, eczema and radiation injuries. Being rich in vitamins it boosts the immune system and prevents colds and flu. It has also been shown to be effective in the treatment of liver cirrhosis.

Link:
http://www.itmonline.org/arts/seabuckthorn.htm

Shepherd's Purse (*capsella bursa pastoris*)

Description: This universal little annual plant is at home almost all over the world and can be found all year round. It's main distinguishing features are the little triangular to heart-shaped seed pods giving the plant their name. They are borne on thin stalks emerging from a rosette of small lobed leaves.

Where: Arable land, waste grounds and gardens all over the world.

When: All year

Culinary uses: All parts of the plant are edible. Young leaves make an excellent addition to the salad bowl although they tend to be a bit small and fiddly. As they grow more mature they increase in spiciness and make a good cress or cabbage substitute as a potherb. The seedpods make a peppery addition to soups and stews. The root is an acceptable ginger substitute.

Medicinal uses: The plant is rich in iron, calcium and Vitamin C. Dried leaves and flowers as a poultice on cuts and bruises. They also are used in the treatment of all kinds of haemorrhages including of the lung, stomach, uterus and kidneys.

Recipe: Vegetable Stew: Gently fry 2 or 3 chopped onions in olive oil. Add 4 diced carrots, 4 diced baby turnips and 4 chopped rosettes of Shepherd's Purse leaves. Cover in stock and simmer for 40 minutes. Add a few seed pods towards the end.

Link:

http://www.pfaf.org/user/Plant.aspx?LatinName=Capsella+bursa-pastoris

Sloe (*prunus spinosa*)

Description: Also known as blackthorn it is an up to 3m tall deciduous shrub which, as the name suggests has a dark smooth bark and some vicious thorns. It is one of the earliest shrubs to flower in spring and the flowers are snow white with 5 petals each. The dark purple fruit is about ½ cm in diameter.

Where: Hedgerows and woodlands all over Europe, although in the south usually at higher elevation.

When: The berries are best picked after the first frost.

Culinary uses: Berries are most famously used in sloe gin. A syrup or preserve can also be made from them. The dried flowers and young leaves make a stimulating tea. Flowers can be crystallized in sugar. Dried berries can also be added to herbal teas.

Medicinal uses: An infusion of the flowers is used in the treatment of diarrhoea, bladder and kidney disorders and stomach weakness.

Caution: All species of the genus contain the toxin hydrogen cyanide, although generally in insufficient concentrations to be harmful. Avoid consuming very bitter fruit, hence it is best after the first frost, which also makes the flavour much more acceptable.

Link:
http://www.sacredearth.com/ethnobotany/foraging/Sloe.php

Sorrel (*rumex acetosa*)

Description: Growing in clumps like spinach. The leaves are slender 5-15cm long, characteristically arrow-shaped at the base and have quite a fleshy texture to them. They have a typical and unmistakable lemon-like acidity when chewed. It displays whorled spikes of reddish-green flowers becoming purplish later and finally bearing brown seeds on up to 60cm tall stalks.

Where: Meadows, along riverbanks and open woodlands all over Europe.

When: Most of the year.

Culinary uses: The French use it in a delicious soup (see recipe). A few young leaves add a lemony bite to a mixed salad.. They can be cooked and pureed and served as a potherb. The roots can be dried and ground to a

powder to add to flour to make bread. The juice of the leaves curdles milk and can be used in place of rennet in cheese production.

Medicinal uses: An infusion of the leaves acts as a coolant for fevers. The leaf juice applied to the skin relieves itching. A decoction of the flowers and/or root in wine is used in the treatment of jaundice and kidney stones.

Caution: The plant, like spinach, contains oxalic acid, which is toxic taken in large quantities. The toxic effect is lessened by briefly blanching the leaves and changing the water. People with tendency to rheumatism, arthritis, gout or hyper-acidity should take special caution not to over-indulge.

Recipe: Sorrel Soup: Sauté 1 large chopped onion, one large handful of sorrel leaves, and 1 handful of chickweed in some butter until soft. Add 1 large cubed potato and 1 litre of hot chicken stock. Simmer for ½ hour until potato is cooked, then liquidise. Add ¼ litre milk or cream and season to taste. Reheat and add an egg yolk in at the last minute. Serve with croutons.

Link:
http://www.botanical.com/botanical/mgmh/s/sorcom64.html

Sorrel, Wood (*oxalis acetosella*)

Description: Not at all related to common sorrel, it does however taste similar due to the presence of oxalic acid. A low growing herb it looks superficially like clover, however the 3 leaf arrangement looks more as if it was made from one single leaf and then cut in three. The flowers look completely different to those of

clover, being white or pink with a delicate hint of purple and 5 distinct petals.

Where: On woodland edges and some meadows, but preferring shady locations all over Europe.

When: Most of the year

Culinary uses: It can be used in much the same way as common sorrel added to salads, soups and sauces. Flowers can be added to salads. The juice is a curdling agent for milk.

Medicinal uses: An infusion of the leaves lowers fevers and is thirst quenching. It is also effective in the treatment of haemorrhages and urinal infections. It is used as a blood purifier and helps calm upset stomachs. Externally crushed leaves can be applied to boils and abscesses.

Caution: The plant like spinach and common sorrel contains oxalic acid, which is toxic taken in large quantities. The toxic effect is lessened by briefly blanching the leaves and changing the water. People with tendency to rheumatism, arthritis, gout or hyper-acidity should take special caution not to over-indulge.

Link:
http://www.botanical.com/botanical/mgmh/s/sorwoo68.html

Strawberry, Wild (*fragaria vesca*)

Description: The wild strawberry looks very much like a smaller version of its cultivated cousin. Groups of 3 small oval, serrated leaves, white, 5-petalled flowers with a yellow centre and finally the pea-sized, bright red fruit, which maybe small, but has so much more flavour intensity than cultivated strawberries.

Where: Sunny slopes on woodland edges.

When: Spring for the leaves, summer for the fruit

Culinary uses: Due to their small size it takes some time to collect a sufficient quantity to do anything but simply eat them there and then. They also bruise quickly if not used immediately. If you manage to gather a decent quantity they make delicious jam or cakes, just like you would use cultivated strawberries. The leaves can be made into a tea and also be eaten (see recipe for spring tart under primrose).

Medicinal uses: An infusion made from dried young leaves and fresh fruit relieve diarrhoea and dysentery. Strawberry juice is effective in the treatment of liver disorders. It is said to lower blood pressure and cholesterol levels. It is also used in the treatment of diabetes, gout, rheumatism and skin diseases.

Caution: The leaves of the cultivated strawberry cannot be used in teas. Some people are allergic to strawberries.

Link:
http://www.liveandfeel.com/medicinalplants/wild_strawberry.html

Strawberry Tree (*arbutus unedo*)

Description: A small evergreen tree up to 5m, occasionally 7-10m tall. The leaves are dark green and glossy, 2-3cm long and slightly serrated around the edge. The small white bell-shaped flowers can often be seen at the same time as the strawberry-like fruit as it takes the fruit a year to ripen. The fruit ripens from yellow to red, is ready when soft and easily picked. The

inside is fleshy and white and the whole fruit has a pithy texture due to its multitude of small seeds.

Where: Woodland edges along the Western Mediterranean and in the far southwest of Ireland (giving rise to its other popular name, Killarney Strawberry)

When: Late autumn, early winter.

Culinary uses: The fruit eaten raw is sweet, but somewhat bland. The Latin name unedo from unum edo – I eat one, suggests that once you've eaten one, you are not really tempted to have another. They are not that bad though. To my mind they are best cooked in jam (see recipe for Christmas Jam under Autumn Olive) or as a liqueur.

Medicinal uses: Not much used in herbalism, but research has shown that all parts of the plant have strong antibiotic properties, which may be useful in the treatment of tuberculosis and leprosy. The leaves, bark and root also act as a renal anti-septic making it effective in treating infections of the waterworks, including cystitis and urinary tract infections. They can also be used to combat diarrhoea and as a gargle for sore throats.

Link:
http://www.pfaf.org/user/Plant.aspx?LatinName=Arbutus+unedo

Sweet Cicely (*myrrhis odorata*)

Description: This herbaceous perennial can grow up to 2m tall with fine fernlike foliage. The white flowers are produced on large umbels.

Where: Hedgerows and woodland edges, it prefers the cooler climates north of the Alps. South of the Alps it is restricted to mountainous areas.

When: Almost all year

Culinary uses: The leaves are used raw or cooked and in small quantities add a sweet aniseed flavour to salads or cooked vegetables. The roots have a similar flavour and can be cooked together with other vegetables. The leaves make a tasty tea.

Medicinal uses: It's diuretic and assists digestion. A decoction of the root is antiseptic and is used to treat dog or snake bites.

Recipe: Apples with Sweet Cicely: Peel, core and slice some sharp apples. Place in a saucepan and add enough water to just cover, plus some sugar to taste. Gently simmer until soft. Add a couple of teaspoons of finely chopped sweet cicely and allow to cool. Eat whole or pureed with a dollop of cream.

Link:
http://www.pfaf.org/user/Plant.aspx?LatinName=Myrrhis+odorata

Thistle, Milk (*silybum marianum*)

Description: This true member of the thistle family must have my favourite Latin name (I'll let you work that one out by yourselves). It is an annual or biennial plant growing to 1.2m tall with shiny, spiky, white-veined, pale green leaves and purple flowers.

Where: On dry wastelands, often close to the sea. Native to southern Europe, but naturalised all over Europe now.

When: Almost all year round.

Culinary uses: Young leaves must have their sharp spikes removed first, after which they can be added to a mixed salad or cooked as a potherb. The stalks peeled and cooked make a celery substitute like cardoons (see recipe). The immature flower heads can be eaten like artichokes. The roots can be boiled and eaten like salsify.

Medicinal uses: The whole plant, but especially the seed, has a beneficial effect on the liver and protects it from damage resulting from alcohol or mushroom poisoning as well as treating all liver related ailments. It is also used in the treatment of gallbladder problems. However the active ingredient, silymarin, is not water soluble and therefore teas made from this plant do not have the benefits associated with bought preparations. Milk thistle also has a reputation to combat depression.

Recipe: Cardi Grantinati: Trim and peel a bunch of thistle stems and immediately drop into some water with lemon juice. Then simmer in salted water for ½ hour until soft. Layer in a buttered baking dish and top with grated Parmesan cheese and melted butter. Season with salt and pepper and finish off with a dollop of single cream. Bake in a very hot oven until brown on top and serve slightly cooled as a side dish.

Link:
http://www.pfaf.org/user/Plant.aspx?LatinName=Silybum+marianum

Thistle, Field Milk or Sow (*sonchus arvensis or s. oleraceus*)

Description: An up to 1.5m tall perennial weed, with dandelion-shaped leaves, although the texture is more succulent and they are framed by some fine spikes, giving the plant the appearance of a thistle. The flowers are yellow and also dandelion-like. When broken any part of the plant exudes a milky latex.

Where: Waste grounds, cultivated fields as a weed, ditches and along rivers.

When: Best in spring.

Culinary uses: The young leaves raw in mixed salads. They are somewhat bitter so should be mixed with less bitter varieties. Cooked as a potherb. The leaves are rich in vitamin C and minerals. It is recommended to remove the marginal prickles before consumption. The young stems can be cooked like asparagus or rhubarb. Young roots can be cooked or roasted and ground as a coffee substitute.

Medicinal uses: Leaves as a poultice are anti-inflammatory. A tea made from the roots relieves the symptoms of asthma, cough and other chest complaints. A tea from the leaves is said to calm the nerves.

Recipe: Sow Thistle Casseroles: Wash ½ litre of sow thistle leaves, don't drain too much, then roughly chop and set aside. Sauté I medium chopped onion in butter until brown. Add the leaves and simmer for 3-5 minutes until soft then season to taste. Add some chopped garlic salami and tip into

114

an ovenproof dish. Cover generously with grated Parmesan and bake at 180C for 15-20 minutes until cheese is melted and slightly brown.

Link:
http://www.naturemanitoba.ca/botany/wildPlants/PerennialSowThistle.pdf

Toadflax, Yellow or Common (*linaria vulgaris*)

Description: Also known as butter-and-eggs, this perennial flower grows up to 90cm high on erect stems with fine, threadlike leaves. The pretty flowers can in some climates appear in mid-winter and are bright yellow with an orange lower lip.

Where: Hedgerows, dry banks, roadsides mostly on chalky soils.

When: Almost all year

Culinary uses: Young shoots cooked or a few flowers in a salad.

Medicinal uses: The whole plant is laxative and diuretic. A tea made from the dried or fresh leaves is beneficial for all liver ailments, gallbladder problems and skin problems. Externally it is applied to haemorrhoids, sores and skin ulcers.

Caution: Some sources report a slight toxicity of the plant. Use in moderation and should be avoided by pregnant women.

Link:
http://www.pfaf.org/user/Plant.aspx?LatinName=Linaria+vulgaris

Thyme, Wild (*thymus*)

Description: There is a bewildering number of sub-species to this genus, in fact there are about 350 of them, many of them varying quite considerably in appearance and aroma. The most commonly found species in Europe are t. pulegioides or broad-leafed thyme, which has broader leaves then other species and a slight lemon aroma, t. serpyllum or creeping thyme, which is very low growing and t. praecox arcticus or wild thyme, which is at home mostly in Western Europe. All are abuzz with bees when their carpet of lilac flowers appear.

Where: All thyme species like open sunny spots on grasslands, heaths, amongst rocks or even dunes.

When: Summer

Culinary uses: Used widely in cooking, as part of herb mixtures, in meat dishes, casseroles, stews and soups. Except for broad-leafed thyme it develops more flavour as it is cooked, but the latter may be used in salads.

Medicinal uses: A tea made from the leaves is used in the treatment of coughs, colds, stomach cramps, colic, poor digestion and loss of appetite. It is also effective against headaches, bowel and bladder problems and as a nerve tonic. An infusion of sage and thyme is gargled against sore throats.

Valerian, Red (*centranthus ruber*)

Description: A perennial growing to 90cm tall with smooth, pointy leaves with a fleshy texture. In the summer it displays a profusion of bright pink flower clusters.

Where: On alkaline soils and along dry stone walls all over the Mediterranean.

When: Spring for the leaves and roots

Culinary uses: The leaves can either be used raw as a salad addition or cooked as a potherb. Opinions are divided on the taste. I like its crunchy, juicy texture, but the aftertaste is somewhat bitter. I only use young leaves in spring in a mixed salad. The roots can be boiled and added to soups.

Medicinal uses: This is not true valerian and has none of its namesakes medicinal properties.

Link:
http://www.plantdatabase.com.au/Centranthus_ruber

Violet, Sweet (*viola odorata*)

Description: Delicate, heart-shaped leaves, slightly downy underneath and the flowers consist of 5 unequal deep purple petals which exude a gentle perfume. They barely lift their heads some 5cm above the leaf cover, but can be spotted from a distance as one of the earliest spots of colour in spring.

Where: Woodland edges.

When: Early spring.

Culinary uses: Leaves and flowers are a good addition to any mixed salad, although the leaves don't have a great deal of flavour to them. Cooked and added to a soup they have a thickening effect. The flowers make pretty cake decorations when crystallised. Leaves and flowers as an ingredient to spring tart (see recipe under primrose).

Medicinal uses: A tea or syrup helps alleviate symptoms of colds, whooping coughs and bronchitis. A decoction or syrup of the roots is used as a laxative. It also contains salicylic acid, a constituent of aspirin and is thus effective against headaches, migraines and insomnia. It is said to have anti-cancer properties.

Recipe: Violet Jelly: Pour some boiling water over 1 litre of fresh violet flowers, stalks removed. Leave to stand over night and strain about ¾ litre of this liquid into a saucepan, add juice of 2 lemons and 1 packet of commercial pectin. Bring to the boil, then add 750g of sugar. Bring back to boil and boil fast for a minute or 2. Pour into hot, sterilised jars and seal. Use on the side with meat dishes or on toast for breakfast.

Link:
http://www.botanical.com/botanical/mgmh/v/vioswe12.html

Walnut (*juglans regia*)

Description: A tall deciduous tree with oval leaves of a slightly leathery texture. The fruit is round and green turning black and dry late in the season indicating that the nuts inside are mature.

Where: Woodlands and parks

When: Mid-June for green walnuts, late summer for mature nuts.

Culinary uses: The mature nuts are a versatile and nutritious food that is used both in sweet and savoury dishes. Great raw in salads such as Waldorf. They combine well with fruit in cakes or as an ice-cream. The unripe fruit picked in mid-June is used to make pickled walnuts.

Medicinal uses: An infusion made from the leaves is used as a poultice against acne, eczema and other skin inflammations. Taken internally it sooths coughs and asthma and eases constipation.

Link:
http://www.pfaf.org/user/Plant.aspx?LatinName=Juglans+regia

Winter Cherry (*physalis alkekengi*)

Description: Also known as bladder cherry, Chinese lantern or Japanese lantern, this perennial grows up to 50cm tall with long, spirally arranged leaves and a very distinctive fruit which is covered by an orange, papery covering looking like a lantern. Only the fully ripe fruit, the size and colour of a cherry is edible.

Where: Cultivated ground and along damp paths in southern Europe. The plant is naturalised in Britain, but probably does not ripen sufficiently to make the fruit tasty.

When: Summer.

Culinary uses: The ripe berries have a sweet and sour taste and can be added raw to salads, including fruit salads. They can be cooked as a jam or candied as a cake decoration.

Medicinal uses: The fruit used fresh, dried or juiced is helpful in the treatment of bladder and kidney diseases as well as gout. Always completely remove the calyx, the papery lantern around the fruit.

Caution: All parts of the plant except the fully ripe fruit are poisonous and should be avoided. Even the ripe fruit should probably be avoided by pregnant women as they have been known to cause abortions taken in quantity.

Link:
http://www.pfaf.org/user/Plant.aspx?LatinName=Physalis+alkekengi

Yarrow (*achillea millefolium*)

Description: An up to 1m tall perennial plant with fine feathery foliage and white to pink umbellate flowers. All parts of the plant exude an intense aromatic aroma reminiscent of chamomile.

Where: Widespread on meadows or lawns.

When: Spring to summer.

Culinary uses: It is really more of a medicinal plant as the flavour is rather intense, but a few sprigs of young leaves can be added to a mixed salad. One report recommends it as a flavouring with fish. An aromatic tea can be made from the leaves and flowers or a few flower heads can be added to flavour a cool drink.

Medicinal uses: Yarrow is mainly known for its blood stemming properties. It can be used both internally and externally as a dressing on wounds, dried and sniffed to stem nose bleeds and as a tea to ease excessive menstrual bleeding. It is also useful for treating colds and fevers as well as kidney diseases. For influenza combine with elderflower and peppermint to make a tea. Apply fresh leaves to an aching tooth to relieve the pain.

Link:
http://botanical.com/botanical/mgmh/y/yarrow02.html

Glossary of Common Names in Various European Languages

English	Latin	French	German	Italian	Spanish
Angelica	*archangelica sylvestris*	angélique	Engelwurz	Angelica	Angelica
Apple, Crab	*malus sylvestris*	Pommier sauvage	Holzapfel	Melo selvatico	Manzano Silvestre
Archangel, yellow	*Lamium galeobdolon*	Ortie Jaune	Goldnessel	Falsa Ortica Gialla	Ortica Muerta Amarilla
Asparagus	*asparagus officinalis*	Asperge	Spargel	Asparago	Espárrago
Autumn Olive Tree	*elaeagnus umbellate*	*elaeagnus umbellate*	*elaeagnus umbellate*	Albero dei Coralli	*elaeagnus umbellate*
Bay Tree	*laurus nobilis*	Laurier Sauce	Lorbeer	Alloro	Laurel / Lauro
Beech Tree	*Fagus sylvatica*	Hêtre	Buche	Faggio	Haya
Beet	*beta vulgaris*	Betterave	Mangold	Barbabietola	Acelga
Bellflower	*campanula*	Campanule	Glockenblume	Campanula	Campanula
Birch	*betula pendula*	Bouleau	Birke	Betulla	Abedul

English	Latin	French	German	Italian	Spanish
Bistort	polygonum bistorta	Bistorte	Schlangen-Knöterich	Bistorta	Bistorta
Bittercress, Hairy	cardamine hirsuta	Cardamine Hérissée	Behaartes Schaumkraut	Crescione Primaticcio	Mastuerzo Menor
Blackberry	rubus fruticosus	Ronce	Brombeere	More / Rovo	Zarzamora
Blueberry / Bilberry	vaccinum myrtillus	Myrtille / Airelle	Heidelbeere	Mirtillo Nero	Mirtilo / Arándano
Borage	borago officinalis	Bourrache	Boretsch	Borraggine	Borraja
Bracken	pteridium aquilinum	Fougère-Aigle / Grande Fougère	Adlerfarn	Felce Aquilina	Helecho Común
Broom, Spanish	spartium junceum	Spartier Jonc	Binsenginster	Ginestra Odorosa	Gayomba / Gallomba
Bryony, Black	tamus communis	Tamier	Schmerwurz	Tamaro	Nueza Negra
Bugle, Blue	ajuga reptans	Bugle Rampante	Kriechender Günsel	Bugola	Búgula
Bugloss, Italian	anchusa azurea	Buglosse Azurée	Italienische Ochsenzunge	Ancusa / Buglossa	-

English	Latin	French	German	Italian	Spanish
Burdock	*arctium lappa*	Bardanne Commune	Große Klette	Bardana Maggiore	Bardana Mayor
Butcher's Broom	*rucus aculeatus*	Fragon	Stechender Mausedorn	Pungitopo	Rusco
Campion, Bladder / Campion, White	*Silene vulgaris / s. latifolia*	Silène Enflée	Leimkraut	Silène / Stridoli / Erba del cucco	Colleja
Caper	*capperis spinosa*	Câprier	Kaper	Cappero	Alcaparro
Carrot, Wild	*daucus carota*	Carotte Sauvage	Wild Möhre	Carota Selvatica	Zanhoria
Cat's Ear	*hypochoeris radicata*	Porcelle	Ferkelkraut	Costolina Giuncolina	Hierba del Halcón
Chamomile	*matricaria recutita / anthemis nobilis*	Camomile	Kamille	Camomilla	Chamomilla
Cherry	*prunus avium / prunus cerasus*	Cerisier	Kirsche / Sauerkirsche	Ciliegio	Cereza
Chestnut, Sweet	*castanea sativa*	Châtaignier	Edelkastanie / Esskastanie	Castagno	Castaño
Chickweed	*stellaria media*	Mouron des Oiseaux	Vogelmiere	Centocchio	Álsine / Hierba Gallinera

English	Latin	French	German	Italian	Spanish
Chicory	*cichorium intybus*	Chicorée	Wegwarte	Cicoria	Achicoria
Chives	*allium schoenoprasum*	Ciboulette	Schnittlauch	Erba Cipollina	Cebollino
Clary	*salvia sclarea*	Sauge Sclarée	Muskatellersalbei	Erba Moscatella	Clary
Clover, red	*trifolium pratense*	Trèfle des Prés	Wiesen-Klee	Trifoglio dei Prati / Trifoglio Rosso	Trébol Rojo
Comfrey	*symphtum officiale*	Consoude	Beinwell	Consolida Maggiore	Consuelda
Coltsfoot	*tussilago farfara*	Tussilage	Huflattich	Tossilaggine	Tusilago
Cornflower	*centaurea cyanus*	Bleuet des Champs	Kornblume	Fiordaliso	Aziano / Azulejo
Corn Salad	*valerianella locusta*	Mâche Sauvage	Feldsalat / Rapunzel	Gallinella Comune	Lechuga di Campo
Costmary / Alecost	*tanacetum balsamita*	Balsamite	Frauenminze / Balsamkraut/ Marienblatt	Erba di San Pietro	-
Cowslip	*primula veris*	Primevère Officinale	Schlüsselblume	Primula Odorosa	Hierba Dentella
Cuckoo Flower	*cardamine pratensis*	Cardamine des Prés	Wiesen-Schaumkraut	Billeri dei Prati	-

English	Latin	French	German	Italian	Spanish
Curry Plant	*helichrysum italicum*	Immortelle d'Italie	Italienische Strohblume	Elicriso	Elicriso
Daisy	*bellis perennis*	Pâquerette	Gänseblümchen	Pratolina	Margarita
Dandelion	*taraxacum officinalis*	Pissenlit	Löwenzahn	Tarassaco	Diente de León
Dock	*Rumex*	Oseille	Ampfer	Romice	Rumex
Echinacea	*echinacea purpurea*	Échinacée	Roter Sonnenhut	Echinacea	Equinácea
Elder	*sambucus nigra*	Sureau	Holunder	Sambuco	Saúco
Fennel	*foeniculum vulgare*	Fenouil	Fenchel	Finocchio	Hinojo
Feverfew	*tanacetum parthenium*	Grand Camomille	Mutterkraut	Partenio	Migranela / Matricaria / Altamisa
Fig	*ficus carica*	Figuier	Feige	Fico	Higuera
Fool's Watercress / European Marshwort	*apium nodiflorum*	Faux Cresson	Kontenblütrige Sellerie	Crescione / Sedano d'Aqua / Erba Canella	Llamada Berra

English	Latin	French	German	Italian	Spanish
French Scorzanera . Bright Eyes	*reichardia picroides / picridium vulgare*	Cousteline / Terre Grêpe / Picridie Vulgaire	Bitterkraut-Reichardie	Lattughino / Cacciapre / Grattalingua / Pizzarello	–
Garlic, Wild / Ramson	*allium ursinum / a. sativum / a. triquetrum / a. roseum*	Ail des Ours / Ail Sauvage	Bärlauch / Wilder Knoblauch	Aglio Orsino / Aglio Selvatico	Ajo de Oso / Ajo Silvestre
Golden Fleece, Smooth	*urospermum dalechampii*	Urosperme de Daléchamps	Weichhaariges Schwefelkörbchen	Boccione / Cicorione / Cicoria Amara	Lechuguilla (Castellan)
Goldenrod	*solidago virgaurea*	Verge d'Or	Goldrute	Verga d'Oro	Vara de Oro
Good King Henry	*chenopodium bonus-henricus*	Chénopode Bon-Henri	Guter Heinrich	Buon Enrico	Zurrón
Goosegrass	*galium aparine*	Gaillet Gratteron	Kletten-Labkraut / Klebkraut	*galium aparine*	Amor de Ortelano / Azotalenguas / Lapa

English	Latin	French	German	Italian	Spanish
Gorse	*ulex europeus*	Ajonc	Stechginster	Ginestrone	Retamo Espinoso
Ground Elder / Bishops Weed / Gout Weed	*aegopodium podagraria*	Égopode	Zaun-Giersch / Geißfuß	Giradina Silvestre	Aegopodio
Ground Ivy	*glechomia hederacea*	Lierre Terrestre	Gundermann	Ellera Terestre Comune	Hiedra Terrestre
Hawkbit, Tuberous *leontodon tuberosus*		Liondent / Leontodon	-	Dente di Leone Tuberoso / Erba Coscia	-
Hawks Beard	*crepis leontodontoides*	Crépide Fausse Dent de Lion	Italienischer Pippau	Radichella	-
Hawthorn	*crataegus monogyna*	Aubépine	Weißdorn	Biancospino	Majuelo / Espino Albar
Hazel	*corylus avallana*	Noisetier	Haselnuss	Nocciolo	Avellano
Heather	*calluna vulgaris*	Callune	Heidekraut	Brugo	Brezo
Herb Bennet / Wood Avens	*geum urbanum*	Benoîte Commune	Nelkenwurz	Erba Benedetta / Cariofillata / Garofania	Hierba Bennet

English	Latin	French	German	Italian	Spanish
Hogweed, Common / Cow Parsnip	*Heracleum sphondylium*	Berce Spondyle	Wiesen-Bärenklau	Panace / Spondilio	Branca Ursino / Espondilio / Pie de Oso
Hops	*humulus lupulus*	Houblon	Hopfen	Luppolo	Lúpolo
Horsetail	*equisetum arvense*	Prêles des Champs	Acker-Schachtelhalm	Equiseto	Equiseto / Cola de Caballo
Hyacinth, Tassel	*muscari comosum*	Muscari à Toupet	Schopfige Traubenhyazinthe	Lampascione	Jacinto Comoso / Hierba del Querer
Jerusalem Artichoke / Sunchoke / Sunroot	*helianthus tuberosus*	Topinambour	Topinambur	Topinambur / Rapa Tedesca	Tupinambor
Judas Tree	*cercis siliquastrum*	Arbre de Judée	Judasbaum	Albero di Giuda	Árbol del Amor
Juniper	*juniperus*	Genévrier	Wacholder	Ginepro	Enebro
Lamb's Quarter / Fat Hen / Goosefoot	*chenopodium album*	Chénopode Blanc	Weißer Gänsefuß	Farinello / Farinaccio	Cenizo
Leek / Wild	*allium rotundum*	Poireau Sauvage	Wilder Lauch	Porro Selvatico	puerro silvestre

English	Latin	French	German	Italian	Spanish
Lemon Balm	*melissa officinalis*	Mélisse	Zitronen-Melisse	Melissa	Toronjil / Melisa / Citronela
Lettuce, Wild	*Lactuca*	Laitue	Wilder Lattich	Lattuga Selvatica	Lechuga Silvestre
Locust, Black	*robinia pseudoacacia*	Robinier Faux-Acacia	Robinie / Falsche Akazie	Robinia Pseudoacacia	Robinia Negra / Falsa Acacia
Mallow	*malva*	Mauve	Malve	Malva	Malva
Marigold, Field	*calendula arvensis*	Souci des Champs	Ackerringelblume	Fiorrancio Selvatico	Maravillo del Campo
Marjoram	*origanum majorana*	Marjolaine	Majoran	Maggiorana	Mejorana / Mayorana
Mint	*mentha*	Menthe	Minze	Menta	Menta
Mugwort / Felon Herb / St. John's Plant	*Artemisia vulgaris*	Armoise Commune	Beifuß	Artemisia Comune	Altamisa
Mullein	*verbascum densiflorum*	Molane	Königskerze	*verbascum densiflorum*	Gordolobo
Mustard, Black	*brassica nigra*	Mostarde Noire	Schwarzer Senf	Senape Nera	Mostaza Negra

English	Latin	French	German	Italian	Spanish
Mustard, Garlic / Jack-by-the-Hedge	*alliaria petiolata*	Alliaire Officinale	Knoblauchsraute	Alliaria	Hierba del Ajo
Myrtle, Common	*myrtus communis*	Myrte Commun	Myrte	Mirto	Mirtos
Naked Weed / Rush Skeletonweed	*chondrilla juncea*	Chondrille à Tige de Jonc / Chicorée à la Bûche	Großer Knorpellattich	Lattugaccio Commune	Achicoria Dulce / Alijungera
Navelwort / Pennywort	*umbilicus rupestris*	Nombis-de-Vénus / Ombilic des Rochers	Felsen-Nabelkraut	Ombelico di Venere	Ombligo di Venus
Nettle, Red Dead	*lamium purpureum*	Lamier Pourpre	Rote Taubnessel	Falsa Ortica Purpurea	Lamio Púrpura
Nettle, Stinging	*urtica dioica*	Ortie	Brennessel	Ortica	Ortiga
Nettle, White Dead	*lamium album*	Ortie Blanche	Weiße Taubnessel	Falsa Ortica Bianca	Ortiga Blanca
Oak	*quercus*	Chêne	Eiche	Quercia	Roble
Old Man's Beard / Traveller's Joy	*clematis vitalba*	Clématide des Haie / Clématide Vigne-Blanche	Gewöhnliche Waldrebe	Vitalba / Erbe dei Cenciosi / Liana	Clemátide

English	Latin	French	German	Italian	Spanish
Orache, Hastate / Orache, Halberd-Leaved	*atriplex hastate*	Arroche Hastée	Pfeilblättrige Melde	Altriplice Commune	-
Oregano	*origanum vulgare*	Origan	Oregano	Origano	Orégano
Ox-Tongue, Bristly	*picris echioides*	Picride Fausse-Vipérine	Natternkopf-Bitterkraut / Wurmlattich	Aspraggine	Lengua de Gato
Parsley, Wild or Cow	*anthrisiscus sylvestris*	Cerfeuil Sauvage	Wiesenkerbel	Prezzemolo Selvatico	Perrifollo Verde
Pellitory of the Wall / Lichwort	*parietaria officinalis*	Pariétaire Officinale	Aufrechtes Glasskraut	Parietaria / Vetriola / Muraiola	Parietaria
Periwinkle, Greater	*vinca major*	Grande Pervenche	Großes Immergrün	Pervinca Maggiore	Hierba Doncella
Pigweed	*amaranthus albus*	Amarante Blanche	Weißer Fuchsschwanz	Amaranto Bianco	Amaranto Blanco / Bledo Blanco
Pine, Stone	*pinus pinea*	Pin Parasol	Pinie / Steinkiefer	Pino Domestico	Pino Piñonero / Pino Albar
Plantain, Buck's Horn	*plantago coronopus*	Plantain Corne de Cerf	Krähenfuß-Wegerich	Erba Minutina / Piantaggine	Hierba Estrella

English	Latin	French	German	Italian	Spanish
Plantain, Common	*plantago major*	Grand Plantain	Breitwegerich	Piantaggine Maggiore	Llantén
Plantain, Ribwort	*plantago lanceolata*	Plantain Lancéolé	Spitzwegerich	Piantaggine / Lingua di Cane	Llanté Menor
Plum	*prunus domestica*	Prunier	Pflaume	Susino	Ciruelo
Polypody	*polypodium vulgare*	Polypode	Tüpfelfarn	Felce Dolce	Polipolio
Poppy, Corn	*papaver rhoeas*	Coquelicot	Klatschmohn	Papavero	Papaver
Primrose	*primula vulgaris*	Primevère	Primel	Primula	Primula
Purslane	*portulaca oleracea*	Pourpier / Porcelan	Portulak	Porcellana	Verdolaga
Raspberry	*rubus idaeus*	Framboise	Himbeere	Lampone	Frambuesa
Rocket, Yellow	*barbarea vulgaris*	barbarée Vulgaire	Winterkresse / Barbarakraut	Erba di Santa Barbara	-
Rocket, White Wall	*diplotaxis erucoïdes*	Fausse Roquette	Rauken-Doppelsame	Ruchetta Violacea	Jaramago Blanco

English	Latin	French	German	Italian	Spanish
Rocket, Wild / Rocket, Perennial Wall	*diplotaxis tenuifolia*	Roquette Sauvage	Schmalblättrige Doppelsame / Wild Rauke	Rucola Selvatica / Rughetta	Rúcula / Rúgula
Rose, Dog	*rosa canina*	Églantier	Hunds-Rose	Rosa Canina	Rosa Silvestre
Rosemary	*rosmarinus officinalis*	Romarin	Rosmarin	Rosmarino	Romero
Saffron	*crocus sativus*	Safran	Safran	Zafferano	Azafrán
Sage, Wild / Clary, Wild	*salvia verbenaca*	Sauge-Vervaine	Eisenkraut-Salbei	Salvia Minore	Gallocreta / Balsamina / Verbenaca
Saint John's Wort	*hypericum perforatum*	Millepertuis	Johanniskraut	Iperico / Erba di San Giovanni	Hipérico / Corazoncillo / Hierba di San Juan
Salad Burnet	*sanguisorba minor*	Pimprenelle	Kleiner Wiesenknopf / Pimpinelle	Sanguisorba / Salvastrella	Pimpinella Menor
Salsify / Goat's Beard	*tragopogon*	Salsifis	Haferwurzel	Barba di Becco	Salsifi
Sea Buckthorn	*hippophae rhamnoides*	Argousier	Sanddorn	Olivello Spinoso	Espino Amarilli

English	Latin	French	German	Italian	Spanish
Shepherd's Purse	*capsella bursa pastoris*	Borse à Pasteur	Hirtentäschelkraut	Borsapastore	Bolsa de Pastor
Sloe / Blackthorn	*prunus spinosa*	Prunellier / Épine Noir	Schlehdorn	Prugnolo	Entrino
Sorrel	*rumex acetosa*	Oseille	Sauerampfer	Acetosa	Acederilla / Acetosella
Sorrel, Wood	*oxalis acetosella*	Oseille des Bois / Alléluia	Waldsauerklee	Acetosella dei Boschi	Acederilla / Alleluia
Strawberry, Wild	*fragaria vesca*	Fraisier des Bois	Walderdbeere	Fragola di Bosco	Fresa Salvaje
Strawberry Tree / Killarney Strawberry	*arbutus unedo*	Arbousier	Erdbeerbaum	Corbezzolo	Madroño
Sweet Cicely	*myrrhis odorata*	Cerfeuil Musqué	Süßdolde / Myrrhenkerbel	Finocchiella	Cerifolio / Perifollo / Mirra
Thistle, Milk	*silybum marianum*	Chardon Marie	Mariendistel	Cardo Mariano	Cardo Mariano
Thistle, Field Milk / Thistle, Sow	*sonchus arvensis / sonchus oleraceus*	Laiteron des Champs / Laiteron Maraîcher	Acker-Gänsedistel / Gemüse-Gämsedistel	Crispigno / Crespigno del Orto	Cerraja

English	Latin	French	German	Italian	Spanish
Toadflax	*linaria vulgaris*	Linaire	Leinkraut	Linaiola	Linaria
Thyme	*thymus*	Thym	Thymian	Timo	Tomillo
Valerian, Red	*centranthus ruber*	Centranthe Rouge	Rote Spornblume	Valeriana Rossa / Lattarola	Hierba de San Jorge
Violet, Sweet	*viola odorata*	Violette Odorante	Duftveilchen	Viola Mammola	Violeta Común / Violeta
Walnut	*juglans regia*	Noyer	Walnuss	Noce	Nogal
Winter Cherry	*physalis alkekengi*	Alkékenge	Wilde Blasenkirsche	Alchechengi	Farolillo Chino
Willow Herb, Rosebay / Fireweed	*epilobium angustifolium*	Épilobe à Feuilles Étroites	Schmalblättriges Weidenröschen	Garofanino Maggiore	Epilobio / Adelfilla
Yarrow	*Achillea millefolium*	Achillée Millefeuille	Schafgarbe	Achillea / Millefoglie	Milenrama

Bibliography

Bonar, Ann, Herbs – A Complete Guide to their Cultivation and Use, 1994, Tiger Books International

Bremnes, Lesley, Herbs, 1994, Dorling Kindersley Handbooks

Culpeper, Nich., Culpeper's Complete Herbal – A book of natural remedies for ancient ills, 1995 (original 1653), Wordsworth Reference

Davies, Jill Rosemary, Hawthorn – Crataegus Monogyna, 2000, Element

Kremer, Bruno P., Wildkräuter und Heilpflanzen – Entdecken und Erkennen, 2011, BLV Buchverlag

Hemphill, John & Rosemary, Herbs – Their Cultivation and Usage, 1983, Blandford Press

Mabey, Richard, Food for Free, 2003, Collins Gem

Marin, Monia, Erbette di Prati e Boschi – Conoscerle e Raccogliere, 2007, Edizioni Gribaudi

Pagliari, Remigio, Erbi –Erbe Mangerecce della Lunigiana e Val di Vara, 2012, Club Alpino Italiano – Sezione Sarzana

•Paume, Marie-Claude, Sauvages et comestibles – Herbes, fleurs & petites salads, 2009, Édisud

Phillips, Roger, Wild Food, 1986, Pan Books

Rangoni, Laura, Liguria in arbanella – Le Buone Ricette Sottovetro della Tradizione, 2007, Microart's

Spoczynska, Joy O. I., The Wildfoods Cookbook, 1985, W.H. Allen & Co.

Treben, Maria, Heilkräuter aus dem Garten Gottes, 1987, Bertelsmann Club

Various, Der große ADAC-Führer durch Wald, Feld und Flur – Natur und Landschaft unserer Heimat, 1978, Verlag Das Beste.

Williams, Jude C., Jude's Herbal Home Remedies – Natural Health, Beauty & Home-Care Secrets, 2003, Llewellyn Publications

Wiseman, John, SAS Survival Guide, 1993, Collins Gem